The Eureka Method

About the Author

Dr. John Hershey holds 144 U.S. patents in fields including spread spectrum, digital TV, 3-D display, medical devices, logistics, e-commerce, jet engine prognostics, radar, cryptography, power line communications, sensors, satellite communications, railroading, and signal processing. He was elected a Fellow of the Institute of Electrical and Electronics Engineers (IEEE) for contributions to secure communications. He is the author or coauthor of eight books and two encyclopedia entries. He worked in the intelligence community and the U.S. Department of Commerce, helped build a regional office of a major government services company, and most recently served at the General Electric Global Research Center in upstate New York. He has taught as an adjunct faculty member at the University of Colorado, the Rensselaer Polytechnic Institute, and the Union Graduate College.

The Eureka Method

How to Think Like an Inventor

John Hershey

McGraw Hill

New York Chicago San Francisco Lisbon
London Madrid Mexico City Milan New Delhi
San Juan Seoul Singapore Sydney Toronto

The McGraw-Hill Companies

Cataloging-in-Publication Data is on file with the Library of Congress

The Eureka Method: How to Think Like an Inventor

1 2 3 4 5 6 7 8 9 0 QFR QFR 1 0 9 8 7 6 5 4 3 2 1

ISBN 978-0-07-177039-2
MHID 0-07-177039-9

Sponsoring Editor
Roger Stewart
Editorial Supervisor
Jody McKenzie
Project Manager
Patricia Wallenburg
Acquisitions Coordinator
Joya Anthony
Copy Editor
Lisa Theobald
Proofreader
Claire Splan

Indexer
Claire Splan
Production Supervisor
George Anderson
Composition
TypeWriting
Art Director, Cover
Jeff Weeks
Cover Designer
Jeff Weeks

Dedicated to Fergus Ross, a good man and a good inventor

Contents

Acknowledgments

I thank Roger Stewart of McGraw-Hill for his mentoring and guidance. His encouragement and patience are deeply appreciated. I also thank Scott Asmus, Rich DeCristofaro, Dave Goldman, and Howard Skaist, four outstanding attorneys, for their taking the time to educate me.

Introduction

This book is designed to help you become an inventor. Books about inventing are already numerous, as are conventions and seminars devoted to the topic. So why, you may ask, does the world need *yet another* book about inventing? The answer is that *The Eureka Method: How to Think Like an Inventor* offers something unique. While most other books focus on what to do with your device *after* it is invented—how to patent it or license it or promote and sell it—*The Eureka Method* steps back to the beginning.

This book will teach you how to *think* like an inventor. In other words, it will teach you how to come up with new ideas for inventions. This is, after all, for most people the hardest part of becoming an inventor.

The fact that you are reading this shows your interest in inventing. In my experience, a person is interested in becoming an inventor for two reasons: dissatisfaction with an existing product or service (too large, too slow, too expensive, too difficult to use), or a dream and desire to create something entirely new, a product or service that will augment humanity's capability to reach farther, move faster, aggregate and analyze all sorts of data, or bring together pieces and form a whole that is greater than the sum of its parts.

Here we examine how you can fulfill both those inventor roles by developing and seizing *white space*. This important term denotes that broad unpainted canvas of inventions waiting to be conceived and is used often in this book. It is the inventor's job to garner, or "color," as much as possible of the white space surrounding an emerging invention by *broadening* and *futurizing*, two techniques discussed in the pages of this book. The inventor can then proceed to seek patent protection on an invention made more valuable by the coloring of the white space in both contemporary breadth and future time.

The Eureka Method Defined

The Eureka Method is a mental discipline that can be learned and practiced to help you produce a Eureka! moment. You may call it an epiphany or a flash of insight, brilliance, or creative genius. It's that moment when an inventive solution finally crystallizes in your imagination. Our name for this critical event is a "Eureka! moment" in reference and tribute to Archimedes who had been wrestling with the problem of certifying a goldsmith's claim that the crown he had made for the king was of pure gold.

The temptation to the artisan was to fashion the crown from a mixture of gold and another element of lesser value, such as silver, and then to charge the royal treasury for a piece of work of pure gold. As the story goes, Archimedes realized, upon stepping into a bath, that he displaced water of essentially the same volume as the displacing object, himself. He also realized that if the displaced liquid were collected, its volume would equal the volume of the displacing object. Since gold is heavier than silver, more silver would be needed to make a crown that weighed as much as gold; therefore, the silver would have greater volume and displace more water.

Upon realizing this, Archimedes supposedly sprang from the bath and exclaimed, "Eureka!" Translated, the Greek word means "I found it!" He had found the solution to the problem in a sudden moment of inspiration. Whether or not this story is actually true, it has nevertheless come to represent, for all time, the moment of inspiration when the scientist, artist, or inventor suddenly and unexpectedly realizes the answer to a vexing problem and this is called the "Eureka! moment."

There is a curious parallel between Archimedes' Eureka! moment and Thomas Edison's instruction to an assistant to calculate the volume of a particular glass bulb having a pear-shaped appearance. The assistant proceeded to use a complex mathematical approach whereupon Edison countered that he would just fill the bulb and measure the volume directly.

History is replete with stories of *Eureka! moments*. Another story, often told in high school chemistry classes, is about German chemist Friedrich August Kekulé, who famously claimed to have realized that the benzene molecule must be shaped like a ring after having a dream of a snake swallowing its own tail. That these breakthrough realizations appear to come unbidden from the subconscious mind makes them seem magical, as if they are beyond our control but, in reality, most of these stories are more legendary than true.

Finding Your Own Eureka! Moments

What is needed is a disciplined approach that can be mastered to help the aspiring inventor reach the point of invention. This book will teach you how to do this and will show you that they emerge from the application of tried-and-true principles rather than flashes of inspiration that come out of nowhere. The application of these principles creates the Eureka! moment of inspiration or realization, which is really the end result of applied reasoning.

Edison, for example, held more than 1000 U.S. patents. He certainly did not wait for inspiration to come from the blue. Like other successful inventors, he applied a methodology to invention. He is quoted as saying, "I find out what the world needs. Then I go and try to invent it." This is certainly a logical first step, which I call "bottom-line-driven invention."

The Eureka Method is a systematic approach to invention that will allow you to find inspiration on a regular basis. The key to thinking like an inventor is to be aware of societal needs and desires, laws and customs. The ideas you are looking for are all around you.

I have found three ways to encourage Eureka! moments. The first is by taking an unbounded thrust at a problem—a sort of no-holds-barred approach that need not be responsive to existing technology or cost considerations. Immersing yourself in a bit of chaos can also help spark creative thought. This can often lead to a "Rube Goldberg" construction. After the inspiration, comes a realization of the constraints of technology and cost, and the initially unwieldy construction can then morph into something novel and useful.

NOTE *For those not familiar with the term "Rube Goldberg," it is defined as "accomplishing by complex means what seemingly could be done simply."*

Another technique for encouraging Eureka! moments is to exploit analogies. As a technical person, you might be aware of an effect, phenomenon, problem, or practice in one field. You and many others might know and understand this effect very well, but suddenly you turn to a completely different set of circumstances and you see a way to exploit an analogy of this effect in this different situation.

A third technique that can lead to a Eureka! moment can be produced by what I call "gaming the system." In this method, you identify the system within which society is operating and then try to find a nonprohibited way around the constraints. This can be a powerful technique, and you have probably used it yourself without realizing it.

In This Book

Chapter 1 presents three questions that you should always keep in the back of your mind when you invent. Answering them for your invention will help you ensure that you are getting the most out of your efforts and will put you in a better position should you seek a patent to protect your invention.

Chapter 2 looks at "improvement inventions." You have probably used a product or system and felt that it could be improved if only there were a component available that did *X*. You might then proceed to invent *X*, or you might search and find that a variant of *X* has already been invented, and with some adjustment it can be made to function and provide the improvement that you envision. Improvement inventions are the lion's share of inventions and a good place to start our journey.

After you feel comfortable with improvement inventions and have learned how to nourish their Eureka! moments, you will want to master the next level of invention complexity: combination inventions. This is where you invent to bring to life entirely new products and services. So that you will have the best chance of success, you should stay aware of important changes not just in technology but also in laws and trends in culture. I have found it helpful to learn how to view a societal system as a game and then to consider an invention as a gaming of that system. You will learn how to game a system as one of the preliminaries to stepping into combination inventions.

Another of the preliminaries to practicing combination inventions is to understand the power that results from combining elements drawn from different sources, the very different technologies that can be combined to produce a new and useful invention. We cast this as an expansion of dimensions, joining two or more spaces of technology and function. You will learn how to seek an opportunity for combination invention using tools such as the technology linkage diagram.

After these preliminaries, we tackle combination inventions. We look at examples that bring together diverse technologies and an example of "the unbounded thrust," and we review the inventive thinking that led to a patented invention by using this approach. We will then study the "POP score," a technique for looking backward and measuring the breakthrough power of a Eureka! moment. You'll then be introduced to a tool to help you generate your own Eureka! moments for capturing combination inventions: the technology linkage diagram.

The final three chapters are dedicated to helping you seek your own Eureka! moments. Chapter 6 covers the importance of laws, regulations, and standards for your opportunities. These paths involve high stakes and potentially high rewards for successful inventors. They are not the usual paths of opportunity, but they can offer some of the best chances to create inventions that have an enormous impact. Technological progress and regulation are closely intertwined. If you can learn to think ahead of the curve, you will be in a privileged position. Remember Edison's statement that he first determined what the world needed and then he proceeded to invent it? If you can foresee what regulations are likely to follow an imminent advance in technology, you would be ahead of Edison's approach and your Eureka! moment might produce an invention that, if protected by patent, would carve out a large and lucrative space.

Chapter 7 deals with examples of overcoming constraints, one of the key motivators for practicing invention. We look at a spectrum of inventions that have resulted from an inventor's push to overcome constraints that typically involve limitations of availability of such things as power and energy; they can also include problems of geography and implicit constraints that emerge from mathematical models.

Chapter 8 is dedicated to what I call "bottom-line invention." This deals with a very important mindset, especially if you are inventing for someone other than yourself, as perhaps invention specifically targeted to a business goal. You will learn how to assume a mindset that is more likely to promote your success by keeping your focus on the business constraints as the invention proceeds.

At the end of each chapter are questions that provoke individual reflection or can be used for group discussion. The questions are designed to help you absorb and consider what you have learned.

Finally, to profit most fully from reading this book, you should be grounded in the essentials of the U.S. patent process. Patents are used to protect the valuable property rights to your inventions, and you should understand and be able to converse in the terminology surrounding patents and the patent process. If you are well versed in patent-related material, it is reasonable to start by reading Chapter 1. But if you are not well versed, or you desire a review or solid introduction to the patent process, you should begin by reading the Appendixes on patents and inventors and inventorship.

A great tool is provided by the U.S. government's website www.uspto.gov. This extraordinarily well-done site is a national treasure of useful knowledge respecting patents, their applications, and a myriad of subjects associated with such. Note that I am not an attorney or a patent agent, and this book does not presume to give you legal advice.

As we have defined it, the Eureka! moment is the instant in which a useful solution to a vexing problem suddenly becomes clear. There is, however, another aspect of the Eureka! moment, and that is its relation to patent law. To gain a patent, you must have invented something useful, novel, and nonobvious. We certainly know what the first two requirements mean, but what about *nonobvious*? That's a difficult one, and case law continues to issue on its definition. Its history goes back to one of the most important moments in the evolution of the patent requirements: In 1941, Supreme Court Associate Justice William O. Douglas spoke for the Court regarding a patent case: "The new device, however useful it may be, must reveal the flash of creative genius, not merely the skill of the calling. If it fails, it has not established its right to a private grant [a patent] on the public domain." This test for nonobviousness lasted only about a decade in the law, but it illuminates the problem that even the highest court in the land has had with defining a Eureka! moment.

Part I

Thinking Like an Inventor

Chapter 1

The Three Questions That Should Follow a Eureka! Moment

If we all did the things we are capable of doing, we would literally astound ourselves.

—Thomas Edison

In this brief opening chapter, I pose three questions that you should always ask yourself when you're considering the implications of your invention ideas. Before you roll up your sleeves and go without sleep or food for days, as good inventors have been known to do, you should first ask and answer these three questions. They are especially important if you plan to protect your invention by securing a patent.

- How can I broaden my invention?
- How can I protect my invention from becoming obsolete?
- Do I understand who benefits from my invention?

I'll also introduce a method that's used throughout this book: examining real-world patents and published patent applications and considering their implications.

My goal is to teach you to think like an inventor. Undoubtedly, you already have a great deal of knowledge about your particular area of expertise, but you are looking for moments of inspiration that lead to original and useful inventions. Getting to those moments, as you shall see, takes careful planning. Throughout the book, I will often quote American inventor Thomas Alva Edison (1847–1931), who was so prolific and whose inventions were so seminal to the modern world that he is widely considered the "patron saint of inventors," if we are allowed to have secular saints. The Edison remark that best pertains here is this famous dictum: "Genius is 1 percent inspiration and 99 percent perspiration." He was probably right about those percentages.

How Can I Broaden My Invention?

A good invention results from your insights into a solution that meets a particular need. To broaden your invention to make use of every opportunity, you need to think about other needs that might be met using your invention, perhaps in a different or slightly modified scenario. When you're considering broadening your invention, ask yourself the following questions: "Where else, and in what forms, can my invention be used? What other problems can be solved by using it?"

Many inventors do not take the time to step back and consider other uses for their invention, and their failure to do this can allow others to capture and exploit opportunities that can be derived, often straightforwardly, from the inventor's creative labors.

Patent Application Preparer

As an inventor, you might attempt to prepare and file for a patent by yourself, but unless you are very familiar with patent law and prosecution procedure, the sometimes long and tortuous negotiation between the inventor and the U.S. Patent and Trademark Office (USPTO), it may behoove you to retain the services of either a patent attorney or a registered patent agent. A roster of patent attorneys and patent agents is maintained for public use at https://oedci.uspto.gov/OEDCI/.

Your patent application preparer can help you broaden your application, but you should always consider that the patent application preparer is quite likely not skilled in your inventive art and the responsibility for broadening will devolve to you. Take a deep breath, put down the soldering iron, and let your mind expand.

Application-Agnostic Inventions

What is meant by the term "application-agnostic"? In the realm of invention, it refers to a technology that has broad impact and provides solutions to many niche problems. An application-agnostic invention's uses will continue to broaden as inventors think of new ways to use the invention to solve particular problems.

In preparing a patent application to protect your inventive rights, you can structure it in a business, or application-agnostic, fashion, so that your patent will have the greatest reasonable breadth. For example, a new turbine engine design for an aircraft might also be used to power another transportation mode, or it might be used as a fixed generator at a terrestrial power plant.

Consider perhaps the most important invention of all time: the wheel. (No, the wheel was not patented.) It is believed that the earliest people to use wheels were the Mesopotamians who, circa 3500 B.C., used a wheel in pottery making. About 300 years later, the Mesopotamians turned the wheel on its edge and used it in chariots. The same invention had tremendous implications for at least two important, and quite distinct, functions.

A more contemporary example of an application-agnostic invention is the radio frequency identification (RFID) tag. RFID technology is often a part of modern logistics support, such as inventory tracking, authentication, and maintenance, and anti-shrinkage support. The RFID tag is essentially a bar code equipped with intelligence and a transponder (a device for sending and receiving radio transmissions). Such a basic and important technology can benefit all sorts of applications, and its use can be greatly broadened.

If you had invented the basic RFID tag, how many truly different applications would you have foreseen? Let's take a look at two applications to get a sense for the broadening potential of such an invention.

Example: RFID and Pet Access

The pet and pet supply business in the United States is a significant industry. Every year, U.S. pet owners spend close to $50 billion on purchasing and maintaining pets with care and feeding. Not surprisingly, a lot of inventions are targeted at pet maintenance. One such invention is taught in U.S. Patent 7458336—an animal identification and entry control system based on an animal collar–mounted RFID tag. The system allows animals to enter a feeding area by approaching an RFID-controlled door or latch: via the collar-mounted tag, the animal triggers a switch or latch mechanism that releases the door or latch. This invention is used in cat feeding areas, for example. As the cat approaches the door, the RFID tag on its collar sends a signal to the component on the cat door latch, and the cat pushes open the door and is allowed inside to eat.

What makes this patent instructive, I believe, is that it demonstrates how a lone inventor can work to enable his vision at his workbench, without having to develop new technology, but instead putting together available components to produce something novel, useful, and nonobvious. Read through just a few lines from his patent summary to see what I mean:

The RFID means used in the system described hereinafter includes a commercially available digital RFID system including the collar tag elements, each with an individual digital code programmed into the internal memory. A typical system of this type is the Texas Instruments Series 2000 Reader System, which is the preferred system for use with the present invention. Other typical control circuitry preferred for use with this invention includes a microcontroller such as the PIC16F876. The techniques of circuit design, antenna design, printed circuit board layout, integration, programming and general use of the RFID circuitry including the commercially available RFID system, microcontroller and other circuitry disclosed herein are well known to those skilled in the art.

The preferred means for RFID also consists of signal processing circuitry for initial processing of the incoming signal, a door control circuit including a sensing switch, a latch and a solenoid with driver circuitry, and the single antenna coil.

The means for allowing and denying access is a lockable door, configured to block an aperture formed as an animal-sized opening in a boundary surface such as a feeding enclosure, fence, wall or larger door. The preferred means for locking and unlocking the door and thus allowing an animal to pass through is a latch mechanism actuated by a solenoid operated by the control circuitry.

Example: RFID Sponges and Surgery

A medical operating room is a busy place, and doctors are equipped with all sorts of individual tools and support items, including sponges for absorbing a patient's body fluids. A somewhat disturbing article in the *Los Angeles Times* (by Shari Roan, October 5, 2010), entitled "No surgical sponge left behind," describes a peculiar problem in a frenetic operating room:

I was observing an operation once when, near the end of the long and tense procedure, a manual count of the surgical sponges showed one was missing. The following few minutes were not fun to watch as the exasperated surgical team went searching for the wayward sponge. Leaving a sponge in a patient, which is easy to do because it soaks up blood and can be hard to see, is not uncommon although the estimates of this mishap range from 1 in 1,000 to 1 in 18,000 operations. Patients who go home with a sponge can suffer later infection and pain.

Now, however, a way to keep track of sponges—square, cotton sheets that are typically 4-by-4 inches or 12-by-12 inches—is under development. Researchers reported Tuesday that they can flag sponges with radio frequency tags or bar codes. In a study presented at the annual meeting of the American College of Surgeons, Dr. Christopher C. Rupp, of the University of North Carolina at Chapel Hill, reviewed 2,961 cases in which sponges with radio frequency tags were used and found the technology helped to recover 21 missing sponges.

The ClearCount Medical Solutions Company stepped up to the challenge, and using RFID technology, it produced a system for counting and detecting surgical sponges. Each new sponge is physically mated with a unique RFID chip. A surgical team member registers each new sponge before its use by passing it over a scanner. Then a display informs the surgical team of the number of registered sponges that are ready for use. After surgery, the used sponges are placed in a bucket equipped with an RFID reader, and the sponge count is displayed. Finally, if the count of sponges in and sponges removed do not match, the team passes a wand over the patient's body that will alert staff if it detects a sponge.

How Can I Protect My Invention from Becoming Obsolete?

Another very important consideration for patent applications for improvement inventions is the concept of "futurizing" the application, and this is a real challenge. A patent's term is 20 years from its filing date, which is a very long time, especially in the world of technology, and also in politics and business: 20 years means at least three different presidents and approximately the "lifetime" of five general managers in some fast-paced companies. As an inventor, you must be very careful about what is claimed and how it is claimed in the patent to ensure that the patent *maintains its value over time*. Not only must you know a bit about the original invention's past, but you must know what's contemporary and where the invention is going, and you must do your best to envision what the future of the invention might be. This is a tall order.

As an important step in patent filing preparation, *futurizing* means that you must consider the alternatives that are presently available or might become available in the future for constructing the improvement invention. Each of the different embodiments of the invention can have many possible incarnations; you should try to cover as many contingencies as possible to fill in the white space as completely as possible. This will help frustrate "design-around activities," which are efforts by other inventors to get-around the bounds of another's patent. It will also make the invention improvement an essential choice for performing the purported tasks.

Foreseeing the Evolution of Memory

Let's explore futurizing with an example that should have a broad resonance. Almost every large system comprises some computation function—that is, something must be calculated. A *utility penalty* is usually associated with the time required to calculate the result—that is, the longer it takes to calculate something, the less desirable it is. We call this the *computational burden*.

In a database for U.S. patents, I entered the phrase "computational burden" and searched for all U.S.-issued patents that included that phrase somewhere in their text. Figure 1-1 displays the number of times the phrase appeared in patents between the first appearance of the phrase in 1974 up through its appearance in 2000. Notice that we never seem to be rid of a computational burden consideration.

Let's take a look at the first patent to use this phrase. It was issued on June 28, 1974, as U.S. Patent 3820894. It protected a system that provided for extremely accurate registration of photo-optical masks. At that time, such a process was becoming increasingly important to the burgeoning integrated circuit industry. The invention provided a control system for a step-and-repeat printing machine. The specification stated the following:

> *The multitude of process steps required to complete an integrated circuit wafer may exceed twenty steps. Many of these process steps are photo-chemical steps, requiring the selective exposure of photoresist through an optical mask,*

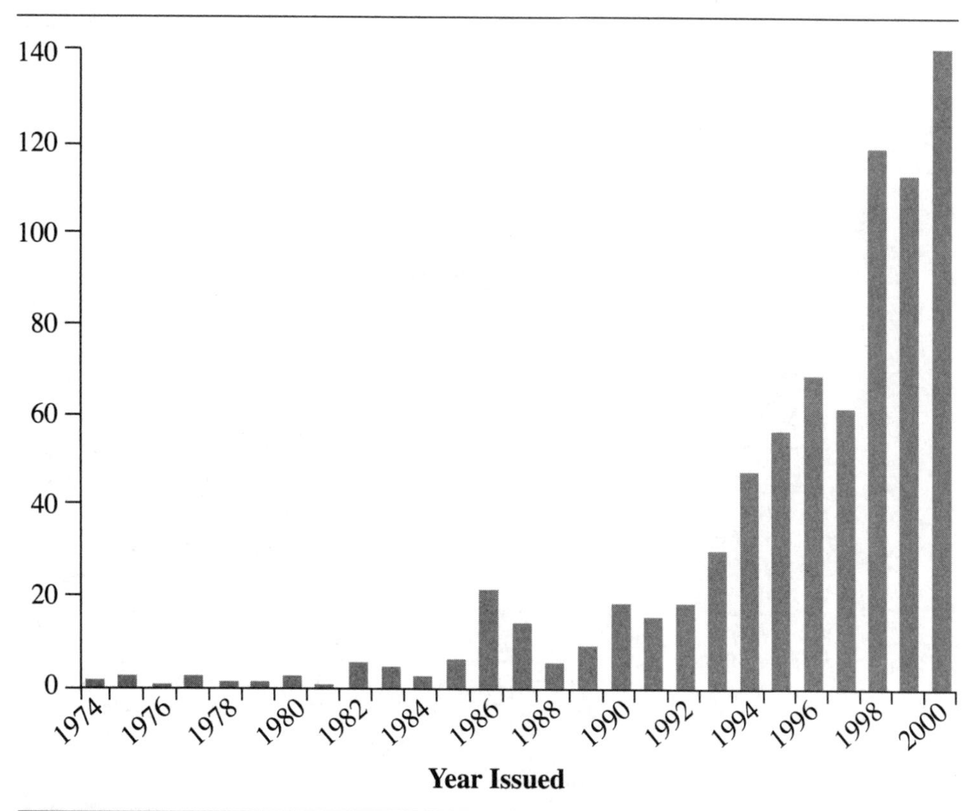

FIGURE 1-1 Number of U.S. patents containing the phrase "computational burden"

then the removal of the masked portion of photoresist to permit selective diffusions or other process operations. It is necessary that these selective operations precisely overlay other preceeding [sic] or following operations to provide controlled die characteristics. The measure of this overlay characteristic is known as registration.

In the manufacture of integrated circuits, registration is critical. Each successive mask must register with all other masks. Corresponding areas of each die must register when the masks are aligned, one on top of the other.

The computational function was required for a variety of system-support functions. The patent's specification included the following: "The present invention more particularly incorporates computer functions of controlling, performing mathematical operations, and storing data into a physically distributed, operatively dispersed system providing control of coaction with extremities."

Now look at Figure 1-2 (the patent's figure 3). It illustrates the integration of the computational function into the overall system. Pay particular attention to the

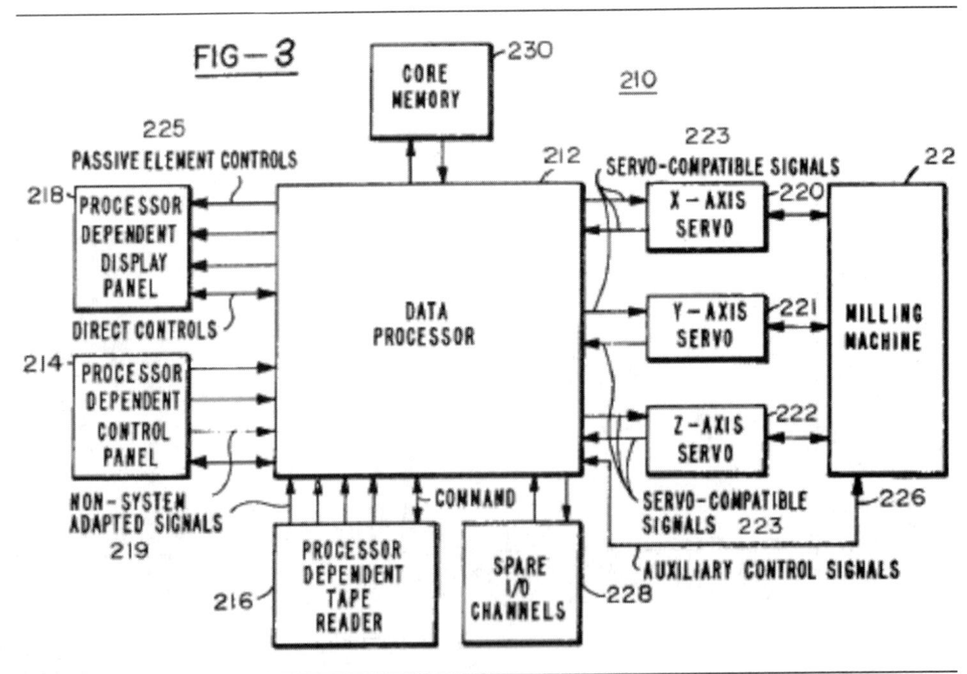

FIGURE 1-2 U.S. Patent 3820894's figure 3

memory module at 230, "Core Memory." The system requires this memory for read-only purposes.

Today, it seems quite possible that most computer scientists and computer designers would not even know what a "core memory" is. They might, however, have heard of a "core dump," a term that is intimately related to this historical memory modality.

For two decades, mainframe computers used core memory, a planar array of small toroids, or "cores," of material that could be set into one of two magnetic states. Each core was thus capable of representing and storing a single bit of information. As is generally the case with most technologies, core production costs declined with time as more efficient manufacturing processes evolved. For cores, the price per bit of core memory behaved as shown in Figure 1-3 for the period 1960–74, declining about 20 percent each year.

What eventually issued as U.S. Patent 3820894 was filed on April 13, 1972, and in that year it didn't make sense to build a computer that did not use a core memory. But, as is the case with most technologies, there were inchoate or even almost mature technologies nipping at the heels of the practiced art. In this case, semiconductor memories were a promising future technology, but it was not until

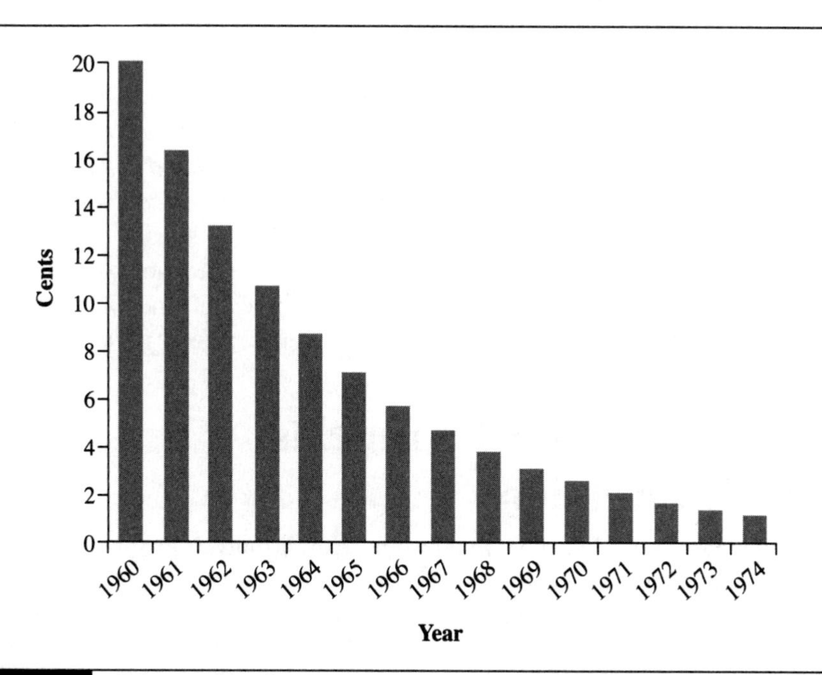

FIGURE 1-3 Price per bit of core memory

1974 that the cost per bit for semiconductor memory dropped to the same price as that of core memory. Of course, it was then only a short time before core memories were essentially abandoned as designers turned to a technology that was cheaper, faster, and required less power to run.

So the inventor of U.S. Patent 3820894 needed to disclose that the best mode for practicing the invention was via magnetic core memory. But with the winds of change palpably blowing, how would the patent be positioned for a longer life?

The inventor tried to meet the challenge by expanding the dimensionality of the memory space technology and by not limiting the patent to a core memory. This was done by adding the following two comments into the specification:

> *In this application, the core memory* **230** *(CM) may be replaced by a read-only memory (ROM) or flip-flop memory....*
>
> *A general purpose data processor is provided which is fully implementable with integrated circuits. Thus, an integrated circuit read-only memory (ROM) provides an example of a capability not found in present data processing systems. Other examples are a random access memory (RAM) and other types of flip-flop memories used alone or in combinations of integrated circuit memories for this data processing system.*
>
> *The data processor* ***212*** *is shown by way of this example to be used in conjunction with a core memory* ***230****. The basic architecture of this data processor will permit an integrated circuit memory, such as a read-only memory (ROM) or a random-access memory (RAM) or flip-flop type memory to be substituted for the core memory* **230** *to provide a completely integrated circuit computer which might be called a monolithic computer.*

The inventor wisely recognized that the enabling technology would soon become obsolete, but by extrapolating likely scenarios based on emerging technologies, the patent itself could be protected from obsolescence, at least for a time.

Do I Understand Who Benefits from My Invention?

Now that sounds like a weird question! How would an inventor not know who would benefit from the invention? It happens. I expect it has happened to you, and it has certainly happened to me, more than once. Sometimes the beneficiary of an invention is not who you expect. As an inventor you should know where and to whom the value of your invention lies. It is not uncommon for an inventor to miss the appropriate audience in selling or promoting an invention.

Let me tell you a story. I started thinking about inventing an improvement for a vending machine after experiencing a common problem with automatic dispensers. I had paid for a package of dried fruit and nuts and pressed D9. The coil rotated 360 degrees and then stopped, but the package held on, hanging from the nearest flight of a horizontal helix. It wouldn't dislodge, not even after a few well-administered knocks to the machine. Then my frustration turned to something positive. Why hasn't somebody invented a simple improvement that should clear a great many of these missed dispensations? It should be relatively easy to cause the motor rotating the helical coil to rotate a bit beyond 360 degrees—say, 450 degrees—and then counter-rotate by the angular amount beyond 360 degrees, which, in this example, would be 90 degrees. This extra rotation should clear up many of the hang-up problems.

Was I the first to think of this? I consulted a patent database search engine, and within an hour I found "my" improvement invention. It is disclosed in published U.S. Patent Application 20070235468 "Programmable Helical Coil Dispensing System."

> *The invention was taught for a system and for the method taught in claim 6:*
>
> *6. A method of dispensing a vending product from a vending machine having a rotatable helical coil to dispense the vending product, the method comprising the steps of: selecting a desired vending product in the vending machine; rotating the helical coil 360 degrees plus an additional angular displacement in a first rotational direction; and rotating the coil the additional angular displacement in a second rotational direction opposite the first rotational direction.*

But let's see what the inventor, Robert Liva, has in the patent application background section to motivate the utility of his invention (bold italics are mine):

> *Cabinet-type vending machines employing helical coils to selectively dispense numerous types of snacks and other goods are well known. It is quite common for vending coils within the dispensing machine to be used in conjunction with a horizontal support tray to dispense vending products. The goods are typically hung from the vending coil. By rotating the vending coil, the item to be vended is longitudinally advanced along the tray, under force from the windings of the vending coil until it reaches the front end of the tray. Upon further rotation of*

> *the vending coil, the item to be vended is forced over the edge of the tray and falls into a discharge bin where the desired product can then be retrieved by the buyer.*
>
> *However, there have been significant problems associated with the helical coil vending machines. Oftentimes, the product to be vended gets caught up on coils and fails to fall into the discharge bin.* ***Besides the failure to properly dispense the product, it is quite common for the user to damage the vending machine in his frustration in attempting to get the selected vended product to dislodge from the coil.*** *A system and method are needed to ensure that goods are properly dispensed from a helical coil vending machine.*

So who would benefit materially from this invention? Certainly not the customer! The customer might be annoyed (and might think about an improvement invention), but there is no immediate direct benefit to the customer. The product line being vended is also not significantly impacted or rewarded by the invention. It's the machine that benefits by incurring less abuse if the improvement invention is adopted. So it is the vending machine manufacturer that might be interested in the rights to such an invention, and for the reason of reducing wear and necessary maintenance on the vending machine—*not* to try altruistically to increase customer satisfaction with the vending process.

Patent Searching

Before you commit to costly research and product development, you might want to know what is already protected by patent or under examination for possible patent protection. Filings are not available to the public until they are published, which is generally about 18 months after filing. But even with this delay, a good search of the U.S. Patent and Trademark Office (USPTO) files may turn up very useful information regarding what others consider worthwhile targets for patent protection. A search is also a good way to review the field.

I consider patent searching more an art than a science, and perhaps the best way to learn is by just doing it. You'll find excellent published resources on searching patents, and you can purchase and subscribe to a number of tools, but if you want to get in the game quickly, the USPTO provides an excellent online search tool that is available to anyone at no cost. If you want to view an application or issued patent along with its figures, you will find the USPTO's tool at http://patft.uspto.gov/netahtml/PTO/patimg.htm very helpful.

Let's consider an example. Suppose that you are interested in the Smart Grid, the biggest infrastructure project in a long time. According to the government's

National Institute of Standards and Technology, or NIST, "Smart Grid will be one of the greatest achievements of the 21st century. By linking information technologies with the electric power grid—to provide 'electricity with a brain'—the Smart Grid promises many benefits, including increased energy efficiency, reduced carbon emissions, and improved power reliability."

The grid's complexity is driven by the distributed nature of the facilities to be coordinated. Forty percent of U.S. energy consumption is for producing electricity. This results in an annual national "electricity bill" of nearly a quarter of a trillion dollars. There are about 10,000 power plants with assets valued at $800 billion. This enormous industrial base serves 131,000,000 electricity consumers. The service structure involves more than 10,000 transmission substations; 300,000 miles of transmission lines; and more than 3000 electrical utility organizations. It is a goldmine of opportunity for the inventor, and the rush to carve out white space is in full swing. One area of intense interest is remote control of home appliances for better power management.

Let's do a simple search to see if we can find any published patent applications concerned with the Smart Grid that protect inventions concerned with "appliance performance" and use of the "Internet."

1. Go to www.uspto.gov/ and click Search under the block labeled "Patents."
2. On the next screen, click USPTO Patent Application Full-Text and Image Database (AppFT), and then click Advanced Search. Notice that all the fields can be searched individually or in Boolean combination.
3. Look for "Smart Grid" in the specification and "appliance performance" and "Internet" in the claims.
4. Following the USPTO instructions, enter **spec/"smart grid" AND aclm/"appliance performance" AND aclm/Internet** in the Query window.
5. We see U.S. Patent application number 20110040785 titled, "SYSTEM AND METHOD TO MONITOR AND MANAGE PERFORMANCE OF APPLIANCES."

If we examine the specification, we do indeed find references to "Smart Grid," and in the first independent claim we see the following:

1. A system to monitor and manage ***appliance performance,*** *said system comprising: a server arranged in communication with one or more appliances* *located within a structure; said server configured to continuously obtain operational data from said appliances; a central database located remote from said server and arranged in communication with said server to receive and process said operational data; and wherein said server alerts a system user upon variation of said operational data from predetermined operating thresholds.*

In a claim ultimately dependent on the first claim, we see this:

5. The system as claimed in claim 4 where said server and said central database are in communication with one another via an ***Internet*** *connection.*

Recap

Remember that your invention resulted from a Eureka! moment that occurred at an instant in time, and time moves on, quite rapidly it seems. As a single inventor, you face tremendous challenges as you try to answer the three questions posed in this chapter, which usually lead to more questions: Where else can your invention be used? What inchoate technologies will likely impact your invention and perhaps allow others to invent around it and thereby nullify any advantage you might have? How do you position your invention so that it will be broadest in application and positioned for future technological shifts? And who are the players in your space—customers, competitors, users, venture capitalists, and others who may profit or lose depending on what happens with your invention? Who benefits from it? Do you?

Discussions and Reflections

- The invention of Teflon was documented as an accident. Research the various uses of Teflon. Were they all foreseen and patented by one entity? In particular, do a little web research to determine Teflon's early use for an armament possessing a microwave target detector. Teflon allows electromagnetic energy to pass through it without significant absorption, and it also has physical strength and good resistance to heat. Next, check out Teflon's use in cookware. What inventors in what countries did these pioneering works with Teflon?
- What is the future of silicon-based electronics? Could the future hold some replacement technologies based on, say, arsenic instead of silicon? Could

vias, electrically conductive passageways in microelectronic packages, be made out of carbon nanotubes rather than copper?

- Many inventors refuse to expend any effort on business issues associated with inventing. They are, in their minds and practice, pure artists. But it may behoove them to spend some time on the economics associated with exploitation of their inventions. Consider that an inventor invents an improvement to, or an entirely new way of, producing a specific nondurable good, perhaps an electrical component that wears out and needs periodic replacement. Suppose further that the inventor's invention will result in a replacement that is far cheaper or lasts far longer than the commercially available device that is supplanted by the invention. Who benefits? Consider the economic considerations of the various parties, including the consumer and the manufacturer of the commercially available devices. Would the manufacturer be interested in purchasing or exclusively licensing the invention for the benefit of the customers? Or would the manufacturer seek to lock up and prohibit the use of the invention by anybody to maintain existing profit margins? Or could a business venture be so structured that the potential increase in market share by exclusive production could offset the loss in individual unit revenue by an increase in volume?

Chapter 2

Improvement Inventions

Look not mournfully into the Past. It comes not back again.
Wisely improve the Present. It is Thine.
Go forth to meet the shadowy Future, without fear, and with a manly heart.

—Longfellow, *Hyperion*

Why can't you leave well enough alone? Experimenters and inventors hear that a lot, and the reason they do is an essential part of the inventor's nature. Inventors respect the past, are excited about the future, but are never satisfied with the present. As the patron saint of invention, Thomas Edison, summed it up this way: "There's a way to do it better—find it."

Improvement inventions are a natural place to start: All you need to do is begin tinkering with an existing device or method sitting before you. How could you make it cheaper, faster, smaller, less power hungry? Or, with a different perspective, how could you improve the process used to manufacture the present device? With virtually unlimited opportunities for creating inventions, your decisions can be guided primarily by your passion for the art and its problems.

An improvement invention can be significantly valuable if it offers an improvement to a popular consumable that sells a large number of units each year. A good change that makes a popular product better can leverage a vast amount of purchasing power away from a competitor's similar product.

To show how improvement inventions are created, this chapter examines some specific examples drawn from the public records of the United States Patent and Trademark Office (USPTO). We'll examine how others have changed existing devices or processes to produce new inventions and new patents that build on existing inventions. The resulting improvements are often both simple and ingenious.

We'll also look at some industries and infrastructures that are ripe for improvement inventions, and I'll profile examples of some of these inventions. You'll learn how to spot opportunities for improvement and how to focus your inventive power to increase the probability of your inventing a meaningful improvement.

"I never did anything worth doing entirely by accident.... Almost none of my inventions were derived in that manner. They were achieved by having trained myself to be analytical and to endure and tolerate hard work."

—Thomas Edison

Building a Better Mousetrap

Only rarely does an invention remain so good that it defies improvement, so that improvements are insufficiently cost-effective, no matter what technological advances have occurred since the original invention. Let's start by taking a quick look at one of these rare examples that defies improvement: William Hooker's mouse trap.

Ralph Waldo Emerson was believed to have said, "Build a better mousetrap, and the world will beat a path to your door." The first mousetrap patented appears to have been that of Hooker in U.S. Patent 528671, "Animal Trap," issued November 6, 1894.

Hooker wrote the following in the patent application:

The object of the present invention is to provide, for catching mice and rats, a simple, inexpensive and afficient [sic] trap adapted not to excite the suspicion of an animal, and capable of being arranged close to a rat-hole, and of being sprung by the animal passing over it when not attracted by the bait.

A further object of the invention is to provide a sensitive trap which may be readily set, and which will be instantly sprung at the slightest attempt of an animal to obtain the bait.

The invention consists in the construction and novel combination and arrangement of parts hereinafter more fully described, illustrated in the accompanying drawings, and pointed out in the claims hereto appended.

Hooker's original conception is discernible from the first figure of his patent, shown in Figure 2-1.

Much has happened in the century since Hooker's invention, but his mousetrap has withstood improvement despite all of the electronic breakthroughs and

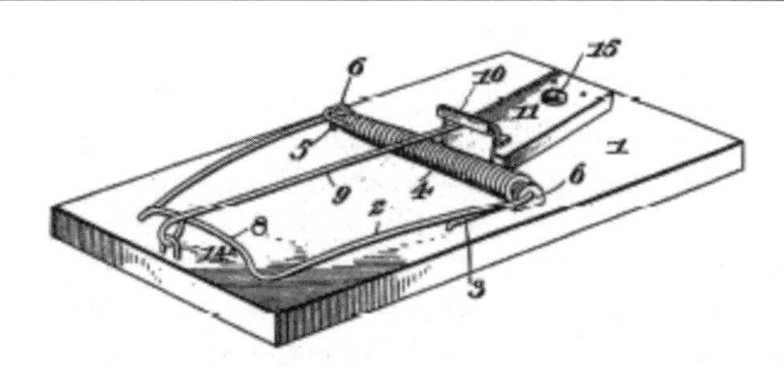

FIGURE 2-1 The once and future mousetrap

material science advances. His invention is inexpensive to produce, simple to understand and operate, and, most important, sufficiently effective. But the mousetrap is an exception. Most inventions, even simple ones, can usually be improved, and the improvements can provide substantial benefits and profits.

How Small Things Add Up

Successful improvement inventions are often innately tied to economics, and one of the most significant economic factors can be scale. In other words, the value or margin of an improvement in a single item may be very small, but if the total number of improved units, the scale, is very large, the improvement effort makes economic sense.

The history of the paper clip holds an excellent set of examples of successful product improvements. The clip type sold as Gem is well known and widely used today. In his book, *Elegant Solutions: Quintessential Technology for a User-Friendly World*, Owen Edwards writes the following:

> *If all that survives of our fatally flawed civilization is the humble paper clip, archaeologists from some galaxy far, far away may give us more credit than we deserve. In our vast catalog of material innovation, no more perfectly conceived object exists.*
>
> *With its bravura loop-within-a-loop design, the clip corrals the most chaotic paper simply by obeying Hooke's law [which states that the extension of a spring is in direct proportion with the load].*

As popular as the Gem paper clip has been and still is, it has not discouraged several successful improvements. In the 1920s, for example, ridged paper clips were introduced, such as taught in G.F. Griffiths' invention under U.S. Patent 1654076, filed April 24, 1923, and issued December 27, 1927. The Griffith patent includes the following:

> *My invention relates to improvements in paper clips of the type known generally as "Gem" clips and it has for its general object to provide a clip in which the portions of the surfaces which are adapted to contact with the sheets of paper or other like material held thereby are roughened so as to more firmly hold and grip such sheets.*
>
> *A further object of the invention is to provide a clip which is adapted to clamp and hold efficiently a plurality of sheets of paper or the like but without marring and without puncturing the same.*

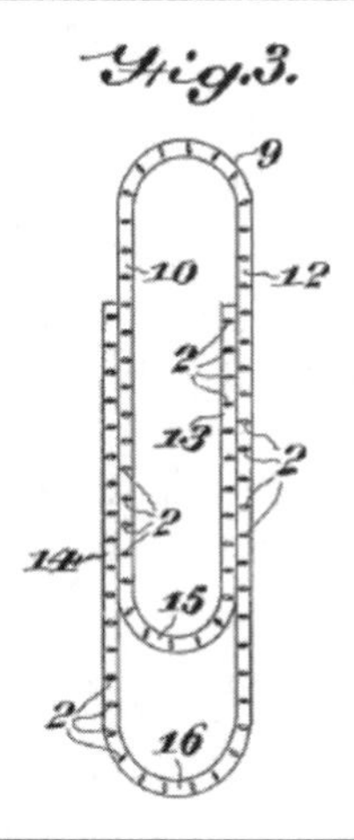

FIGURE 2-2 The ridged paperclip

The patent's third figure, shown in Figure 2-2, should look quite familiar.

This demonstrates that the simple paper clip was indeed the source of a successful improvement invention. In this case, the single *dimension of improvement* is to the clip's ability to grasp or hold together "a plurality of sheets of paper." You can tell that inventor Griffith was a good experimenter by the way he described fabricating the notched paper clip wire: "I have found that they [the notches] may be most economically and efficiently produced by passing the wire between a couple of oppositely disposed and co-acting knurled rollers."

The Dimensions of Invention

I find it useful to think of the inventive process as what I call the "dimensions of invention." The inventive process is not limited to the "usual" dimension considerations: height, length, width, and perhaps time. Invention dimensions comprise such factors as weight, power consumption, speed, basic technology, or just about anything that is a feature of the device or method that you're seeking to improve. In designing an improvement invention, you generally focus your inventive effort on a single dimension.

As you progress through this book, you'll examine efforts directed toward improving more than one dimension, and you'll read about inventions that resulted from increasing the number of dimensions by combining components, subsystems, and features of different technologies and smaller dimensional inventions.

Using Analogy to Help You Improve Inventions

It might prove helpful to have or get some familiarity with the theory underlying the technology in which you choose to invent. One of the most compelling reasons to do this is to empower you to "harvest" analogies, or similarities, from related arts, products, and inventions. Comparing technologies and related arts can often seed a Eureka! moment, just as Archimedes recognized the analog between displacing water and measuring the volume of an irregular object. As you think about similarities between technologies, you may come across ways to use a technology, or part of it, for another application.

Microwave Ovens and Cell Towers

For an example of an invention that improves a product's manufacturing process, let's consider the microwave oven, an appliance we will revisit later in the book.

In its basic form, a microwave oven is a closed electrically conducting box or cavity into which microwave energy is pumped. Typically, the microwave radiation is at about 2.45 GHz and has a wavelength of just less than 5 inches. Food that is placed within the oven is heated by the electromagnetic energy passing through it and vibrating the water molecules, causing heat to be generated by friction. The more energy present in a volume element of the food, the faster that volume element will be heated.

Basic microwave ovens cook unevenly, with cold spots found within the food. These cold spots result from uneven heating that results from an uneven distribution of microwave energy within the oven. But why is there an uneven distribution of energy? Understanding this question and its answer is crucial to understanding the progress made in microwave oven design and progress that has yet to be made.

As mentioned, a basic microwave oven is a closed box with conducting walls. Electromagnetic microwaves will not pass through an electrically conducting wall; instead, they are reflected by the wall. They are also scattered by conductors or partial conductors, such as food items, that are inside the oven. You can think of the microwave radiation within the oven as a great number of microwave rays that are being reflected and scattered, and as a result, the energy at any one point depends on the direction and phase of the many different rays that are coincident at that one point at any one time. The energy will therefore vary greatly throughout the oven's interior.

To think like an inventor you seek to draw upon analogies within related fields. Where else does a similar situation and such an effect occur? Where else do we find "hot spots" and "cold spots"? The answer is well known to any cell phone user. You can receive an adequate signal in one spot, but move just slightly and the connection may severely degrade. Or, if you are standing next to a busy street, the signal can suddenly fade and just as suddenly return.

These issues occur for reasons similar to the microwave problems. A microwave signal is transmitted from a cell tower and reflects off many objects in the environment, resulting in rays arriving at the handheld cell phone from a variety of locations. Sometimes the individual rays add and strengthen the signal, but sometimes they produce cancellation. In the cell phone world, this phenomenon is known as *multipath fading*, in which the signal can fade significantly as a result of just a small spatial displacement of the cell phone; this fading is further classified as being *fast*.

The cellular situation differs from that of the microwave oven in a number of ways. One very important difference is that moving and changing reflectors influence the addition and subtraction of the individual rays received by a cell phone. These moving or changing reflections can be caused by the microwave signal being deflected and reflected by automobiles, buses, planes, and even people. In the microwave oven, not much changes as far as the reflectors and scatterers, so the hot and cold spots tend to remain static. Techniques used to mitigate this problem of static reflections involve physically moving the food as it cooks or using a technique called *mode stirring*, as taught in U.S. Patent Application 20100243646, "Method and Apparatus for Mode Stirring in a Microwave Oven":

> *Food is thus often turned or otherwise moved physically in microwave ovens. One other technique that may be used to reduce these effects of a multipath-induced heating deficiency is referred to as mode stirring. This technique can be performed in a variety of ways such as through incorporation of a moving reflector near the point where wave patterns are emitted. The moving reflector changes the standing wave patterns and spatially perturbs the nulls in the wave patterns. Mechanical mode stirring arrangements may, however, include a costly and noisy mechanical apparatus to drive the reflector. This mechanical system may entail extra manufacturing time and components in addition to moving parts that may require maintenance later in the life of the microwave oven.*

So the question for the clever inventor is this: How can I provide mode stirring without incurring a costly and noisy mechanical apparatus?

Unfortunately, the path to a worthwhile appliance invention is not always as clear as you might think. The problem is that the cost of producing an improvement, no matter how miniscule, often outweighs the benefits of profits.

It is in fact a standing joke that "Millicent" is not a person who works at an appliance production facility, but rather that small cost edge that can determine competitive price or a substantial margin when multiplied by tens of millions of units.

Scattering the Waves

An invention taught in U.S. Patent Application 20100243646 actually shows a way to achieve mode stirring without adding conventional moving parts, thereby keeping costs within an acceptable margin. The inventors' concept was to draw on the analogy with cell phones and introduce "one or more scatterers with a variable or changing radar cross section, configured to perturb the electromagnetic waves and their patterns within the [microwave oven] cavity." These special scatterers are easily designed and of small size considering the microwave oven's radiation wavelength. The power required to run them can be harvested from the power within the microwave oven's cavity. In addition, the scatterers can be housed not only within the microwave oven's cavity, but also in a removable tray or in specially designed cookware.

Appliance production is an industry that has continually trimmed its operating costs through improvement inventions. Small improvements resulting in lower parts cost or reduction in product production "turn time" can reward the manufacturer with significant profits because of the number of items produced.

Persistent Improvement in the Ski Industry

The ski business is rich with continuing improvement and has provided tremendous and continuing opportunity for improved invention. Because the improvements continue year after year and at a significant rate, the ski business is an example of an art that is "invention-persistent."

The patenting of ski and ski-related technology has proceeded steadily. Figure 2-3 plots the number of U.S. patents relating to skis as a function of filing year, 1990–2000. Notice that continual improvements are made throughout these years, with no significant diminishment of the rate of improvements. This is a result of a number of factors, including the popularity of the sport, with at least 15 million skiers in the United States alone; the discovery and refinement of new materials for making skis lighter and stronger; and the fact that skiers generally fall into the higher economic classes and gladly appropriate discretionary income to the latest and greatest in their sport's technology.

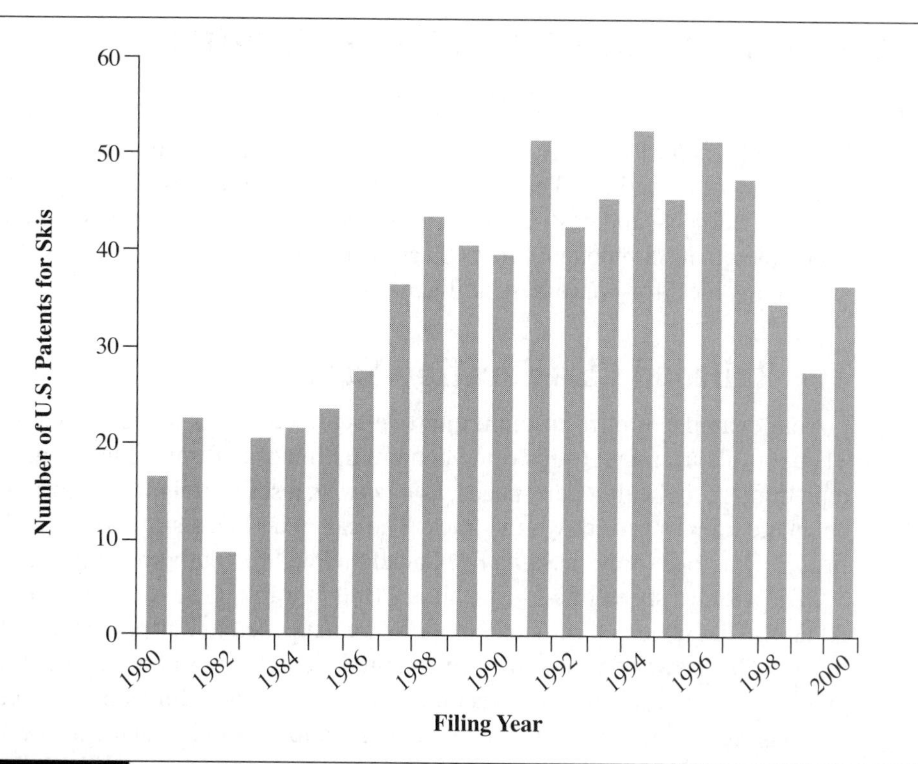

FIGURE 2-3 The number of U.S. patents relating to skis as a function of filing year

The dimensions or opportunities for improvement of the ski itself comprise, at a minimum, length, stiffness, bindings, weight, and cost. For just a sampling of the mileposts in the ski world, consider these improvements:

- 1933: Three-layer patented laminated ski construction
- 1949: First commercially successful aluminum ski
- 1959: First successful plastic fiberglass skis invented
- 1984: Some patenting of lightweight skis
- 1995: First "smart skis" produced

Improvement Invention Opportunities in Infrastructure Upgrades

Large and complex infrastructures are a veritable agar, or culture medium, for improvement inventions. The reason for this is simple: Infrastructure projects are costly, ponderous, and destined for a long history. From the moment a project is completed, its components start to age, while technology marches on. This situation is ideal for improvement invention.

The Railroad Classification Yard

For our example of infrastructure improvement, we'll look at the railroads. Thoughts about freight trains often conjure up images of trains highballing at 70 mph through the wide-open spaces, as a *consist*, a set of freight cars forming a complete train, speeds down the track. The cars read like a storybook: the Katy line, Burlington Northern Santa Fe, Canadian Pacific, Southern Railway.

The average speed of a cross-country freight train, however, is actually about 20 mph. That seems inordinately slow. What happened to our preconceived image? The answer lies in some fundamental elements of rail transportation. In practice, innumerable delays affect the train's speed, including stops for fuel and crew changes, congestion, and also reclassification (as individual cars are split from one consist and added to another). This latter consideration is both basic and fundamental and has a profound effect on a train's average speed.

Fifty, sixty, seventy, or more cars go by loaded with, well, who knows what. Are they all going to the same place? No. So how, then, are they individually dispatched? Let's consider an analogy. In today's computer network parlance, *routers* are the electronic devices that send individual messages to their respective addresses. Think of a message arriving in your company's e-mail inbox as a long string of envelopes, or packets, each needing to be switched out of the message into separate delivery routes or channels. This is similar to how a train is reclassified. In this case, the envelopes are the freight cars. A chief difference, of course, is in the dimension of weight.

Trains enter a classification yard, where they are disassembled and new individual consists are built up from the disassembled trains. Classification yards are complex, large sets of rails and switches, such as the CNW yard in Chicago, shown in Figure 2-4.

One old and intriguing type of classification yard is the "hump" yard. It is a fascinating place, a veritable frenetic microcosm of a railroad system, where incoming trains are broken down and outgoing trains are formed from their cars

FIGURE 2-4 The CNW classification yard in Chicago

using gravity. Cars are pushed up a small hill, a hump, and allowed to run down the other side, where they are switched onto different tracks and form new trains that will take their cars to their destinations or to the next classification yard. Figure 2-5 shows the view looking down the hump of a Chicago hump yard.

After the new trains are assembled, they are sent to the departure yard. The tracks in the departure yard are physically quite close, as you can see in Figure 2-6, which features the author and a colleague, Dr. Ken Welles.

A hump yard is busy and also extremely, painfully noisy. The noise is not predominantly from the rolling stock; actually a rolling car is ideally rather quiet. The noise comes from the retarders. And what are they?

A car rolling down a hump hopefully has enough speed, typically about 4 mph, to reach the train to which it is to be joined and effectively actuate the coupling. The car should not be traveling too fast by the time it arrives at the new train. Too fast an encounter can damage the couplers and also imperil cargo. Humps are

FIGURE 2-5 A view of a hump yard

FIGURE 2-6 Tightly packed trains in a departure yard

configured so that under most conditions, a car will reach more than enough speed to arrive at the new train; *retarders* limit or reduce the speed of the rolling stock to avoid damage.

A hump yard retarder is shown in Figure 2-7. The device applies pressure to two plates on either side of the wheels—the more pressure, the more retarding force (and noise). Retarders are a key component of humpyards and have attracted improvement invention as we shall see.

Retarders have been around for a very long time. An online history of British railroads (http://myweb.tiscali.co.uk/gansg/8-yards/y-marsh.htm) gives a fine account of early hump yard days:

> *Operating a hump yard involved three men at the crest of the hump, one tally man keeping track of the wagons [railroad cars], a shunter with a pole who uncoupled the wagons at the top of the hump and a man who applied the hand brake so the wagon would not roll too quickly down the sidings. The points were controlled from a central signal box, usually called the control tower, but this system still required men to run alongside the wagons and apply the hand brakes in the sidings with the consequent risk of injury. One way to reduce*

FIGURE 2-7 A hump yard retarder

the risk to staff was to supply chocks that could be placed on the track to stop a wagon but these sometimes derailed the wagons. In the 1920's 'automatic retarders' were developed, these were hydraulic chocks that slowed the wagons down to minimize the shock when they hit the already parked stock.

Spotting Opportunities in the Classification Yard

The classification/hump yard is an example of an embedded infrastructure that will not be outdated anytime soon; this makes it an ideal candidate for spotting opportunities for improvement inventions.

There are many opportunities to improve efficiency and to add value through single-dimension improvement inventions. There is little doubt that improvement invention has been at least partially responsible for the remarkable fact that the number of U.S. railroad workers in 2008 was about a quarter of what it was in 1967.

Let's look at four notable areas of improvements, visible through patent evidence:

- **Retarder speed control** The speed of the rolling stock as it proceeds down the hump must be controlled. Too little speed, and the car string will stop short of the train to which it is to be joined. The humping operation will then have to be halted as a service locomotive is brought in to join, or "trim," the cars that have stopped short. Too much speed, and the cars can impact and damage themselves or the train under construction. In some hump yards, even today, this speed is controlled by a seasoned eye and a seasoned guess. But why not import a bit of technology? That is exactly what was done in U.S. Patent 5388525 "Railway Car Retarder," filed on August 19, 1993, which concerns radar-based sensors.

  Claim 19 of this patent states simply:

  A car retarder as recited in claim 18, wherein: said sensors are radar based sensors.

  This was a classic case of improvement invention where a relatively new technology (radar) was combined with a relatively old technology (the retarder) to produce a valuable improvement.

- **Scheduling** When you are dealing with hundreds of railroad cars that can be shuffled in myriad ways, the combinatorics can be mind-boggling and even computer-boggling. The yardmaster or operator of a hump yard is tasked with organizing and orchestrating operations within the yard so that flow is maintained and efficient. Consider U.S. Patent 4883245, filed July 16, 1987. This patent motivates its utility by the following text of the background section (incidentally, large dimensional matrix inversion is an example of a computational burden, which was discussed in Chapter 1 regarding futurizing a patent):

> *The nominally optimal solution to running trains and blocking cars in order to minimize the number of trains operated for a given amount of traffic (and thereby minimize the number of crews and engines used) is given by the integer programming model.... The two major assumptions justifying this model are reasonable: that variable costs are a stepwise function of the number of crews used, with other operating costs for a given amount of traffic being fixed, and that the arrival rates of cars into the system are predictable. However, this model is not commercially viable for two primary reasons: even with the selective elimination of improbable variables, the matrix inversion required to solve this model is too large for available computers.*

Again, thinking like an inventor involves surveying contemporary art and profitably applying it to an old infrastructure.

- **Remote control of locomotives** With the frenetic growth of miniaturized electronics in the 1950s, remote control became a reality. People were flying model airplanes, controlling model boats, and, of course, running model cars in parking lots. Remote control also became a necessity in classification yards. In a classification yard, a *hostler* refers both to the light engine used to move cars around within the yard and to the operator of the engine. Before remote control, a hostler needed to be on board to move a train within the yard. Today, remote control technology is the norm. Let's look at the first claim of U.S. Patent 6856865 filed January 7, 2004:

  > *A portable remote control locomotive device comprising: an operator input for generating locomotive commands; a transceiver for transmitting the locomotive commands to a locomotive; a display; a data base of at least a track profile stored on the device; and a program on the device for determining and driving the display to show the location of the locomotive on the track.*

- **Ancillary wireless communications** What about the myriad other functions within a classification yard? Surely time, effort, and money could be saved by importing and applying widely known and used communications technology. Indeed it can. Consider claims 5 and 6 of U.S. Patent 7257471, filed November 4, 2005:

> *5. A communications device for routing a locomotive over a track layout in a train yard, said locomotive operable over a plurality of alternative track routes to reach a respective destination from a plurality of possible destinations in said track layout, said track layout including a plurality of switches configured to alter a route for a locomotive running along said track layout, said communications device comprising: a user display positioned wayside in the train yard to enable an operator at the train yard with remote control means for controlling movement of said locomotive to command a desired destination for the locomotive within said track layout by setting the state of the switches along the route to the destination through a wireless communications link, wherein the user display is accessible by the operator at the train yard, thereby enabling said operator to individually (1) control movement of said locomotive, and (2) directly set the state of the switches along the route to the destination through the wireless communications link independent from a train yard control system.*
>
> *6. The communications device of claim 5 wherein said communications link comprises a direct communications link between said user display and the plurality of switches.*

The following article about the CSX Corporation, from the August 6, 2008, issue of *Railway Age*, paints an excellent picture of the march of improvement invention in classification yards. You don't have to know or decode the product names, because the flavor comes through quite clearly.

The RailComm-supplied Shove Track control system at CSX's Hamlet Yard in North Carolina will be upgraded with a RailComm DOC (Domain Operations Controller) Yard Automation System. The PC-based control system will provide wireless remote control from the yard tower to all equipped switch locations.

The DOC System features eNtrance and eXit (NX) routing and stacked route planning capability, providing the tower operator with complete control of all routes within the yard. RailComm Switch Position Indicator (SPI) units will be installed at each switch. These will flash when a route is lined over a switch, will be constantly lit when a route is not available, and will flash red when a switch is out of correspondence. A sensor is included to adjust LED brightness to accommodate day and night modes. RailComm's 2.4 GHz RADiANT data radios will provide a wireless communications network to link the office with the field locations.

To be granted a U.S. patent, an invention must have utility—that is, it must be useful. For a discussion of this and other requirements of a patent, see Appendix A.

The railroad hump yard has hosted many different improvement inventions with a variety of applications. Such venerable infrastructures are generally excellent breeding grounds for improvement inventions because of the very nature of the infrastructure—that is, the technology is in place after the infrastructure is built and the technology cannot evolve on its own.

Traffic Lights

In the United States, about 50 million traffic lights are in place at intersections, crosswalks, train crossings, and other roadway features to control the flow of traffic. This infrastructure is being gradually improved by the replacement of the incandescent light bulbs in traffic lights with arrays of LEDs, the substitution of a new technology of lighting for an older one, as shown in Figure 2-8.

By mid-2006, 60 U.S. patents and 47 published U.S. patent applications addressed LEDs and traffic signals. Why so much interest in what seems to be a simple, natural technological progression? The answer derives from the impressive number of utility advantages that are positively impacted by the improvement invention of substituting LED technology for incandescent light technology in traffic signals. Consider the following:

- Operating LED traffic signals requires less electrical power, on the order of a fifth of that required to run an incandescent signal.

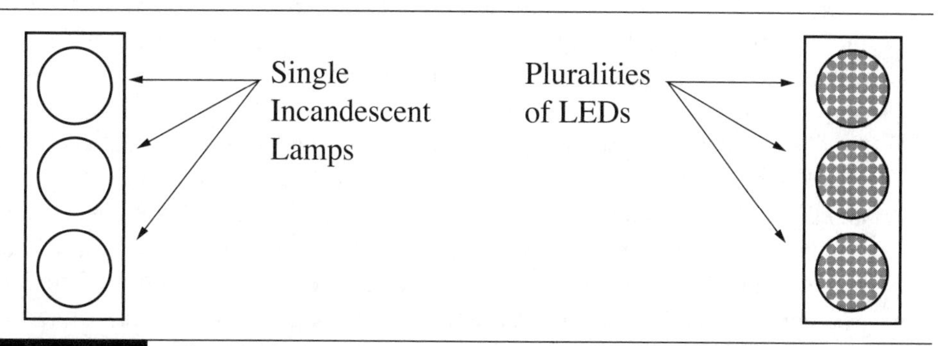

FIGURE 2-8 The substitution of a new lighting technology for an old one in traffic signals

- LED signals are brighter than the replaced incandescent bulbs.
- LED signals require less maintenance than incandescent signals.

Replacing incandescent bulbs by arrays of LEDs in traffic lights is a large undertaking. This particular improvement has some unexpected consequences and unexpected opportunities for a combination invention. We will examine these in Chapter 6 when we look at the role that laws, regulations, and standards can provide in stimulating invention.

Focusing Your Attention Where It Counts

As mentioned, improvement inventions are the lion's share of invention, especially at a company's research and development (R&D) facility, where invention is to some extent a social activity, performed in teams and clearly focused on the company's needs. Without direction, most inventive people will continue to invent: This is their etiology, and without guidance or focus they are likely to produce all sorts of quaint, clever, and largely useless inventions.

To guide the inventive activities of an R&D team, to focus it and make it efficient, it is tempting to rank a system or product's components with respect to cost, weight, power consumption, speed, or whatever dimension is considered fair game for improvement, and then focus on the single dimension that will yield the greatest value on improvement. For example, portable phone size was, at one point, an extremely important marketing consideration. Customers seemed to desire ever smaller phones. For some other devices, particularly in the power generation field, reduction of weight was the single biggest concern.

Suppose, for example, that we desire to focus on reducing the total weight of a system comprising four modules, A, B, C, and D, weighing, respectively, 9 kg, 6 kg, 2 kg, and 1 kg, as illustrated in Figure 2-9. The weight of the system, the sum of the weights of the four components, is 18 kg. Assume that we want to reduce the weight by as much as possible and that we can only afford to focus on one module.

Clearly, the heaviest module is module A. If we focus on the largest weight component, module A, we will realize a weight reduction of at most 50 percent and that only if we are able to reduce the weight of module A to 0 kg. This analysis seems quite straightforward and intuitive. But the analysis gets a bit tricky and counterintuitive when we seek to improve inversely related figures of merit such as computational speed, for example, by improving execution times of various modules. The inverse relation being that computational speed is inversely related

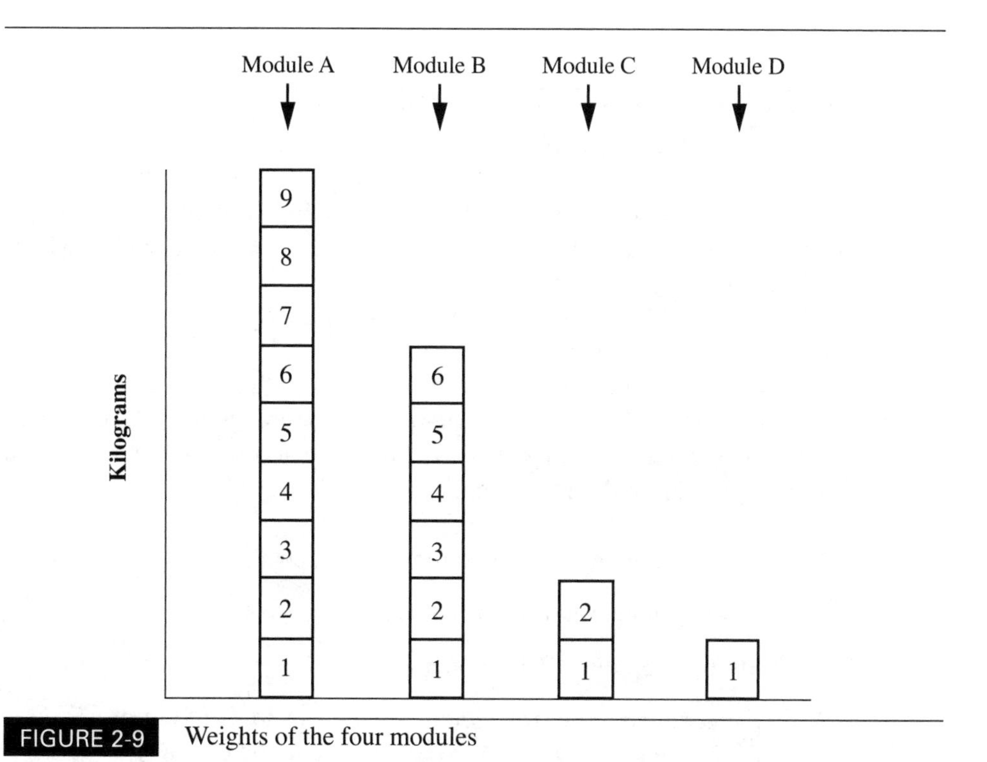

FIGURE 2-9 Weights of the four modules

to execution time. An example of this is provided in the box for those who are interested in just a bit of math.

Selectivity and Moore's Law

The previous examples highlighted the need to model or calculate the greatest improvement that you as an improvement inventor can expect by concentrating improvement on a single component or dimension. But if you are going to invent an improvement against a barrier, shouldn't you start with the biggest one? That just seems to make sense. The answer, however, is "not necessarily" and the reason is that many others might be working on the same problem, and by the time you have expended your effort and your R&D dollars, the industry as a whole may have already beaten the problem.

Suppose, for example, that you want to build and manufacture a small robotic device to pick up trash, and you are planning on a couple of years of R&D before you introduce your robot. You envision that the robot will have a computer, a

A Figure of Merit Involving Inverses

For example, one very common seduction that ensnares many new computer engineers pertains to algorithmic execution speed. (This example adapted from "The Measurement of Performance on a Highly Parallel System," by D. Parkinson and H. Liddell. *IEEE Transactions on Computers* C-32[1], pp. 32–37, 1983.)

Suppose we have two equal-length, N-long strings of non-integer or floating-point numbers denoted by X(1), X(2), …, X(N) and Y(1), Y(2), …, Y(N) and we want to form the N products Z(1) = X(1)Y(1), Z(2) = X(2)Y(2), …, Z(N) = X(N)Y(N). The time required to go through the following loop N times can be easily calculated by first listing the necessary steps and the times required for their individual completions:

1. Get two operands out of memory: X(i) and Y(i).
2. Floating-point multiply them together.
3. Store the product, Z(i), in memory.
4. Increment an initialized counter that keeps track of our progress through the loop: $i \leftarrow i + 1$ (adding one to the counter).
5. Test the counter to see if the loop has been accomplished N times.

For those who remember FORTRAN notation, our loop would look like this:

```
DO 1 i = 1,N
1 Z(i) = X(i) * Y(i)
```

We represent the operational times as:

- Tc = the (cycle) time to put a result into memory or get an operand out of memory
- Ti = the time required for the integer arithmetic to increment the counter
- TL = the time required for testing the counter

- Tfm = the time required for a floating-point multiplication

Using this information, we can now calculate the time, T, spent in one passage through the loop as:

T = 3Tc + Ti + TL + Tfm

Now suppose that Tc = 200 nanosecond (ns), Ti = 100 ns, TL = 100 ns, and Tfm = 1200 ns. For these values, the time needed to execute one passage through the loop is 2000 ns. The computational times ranked by decreasing magnitude are depicted in Figure 2-10.

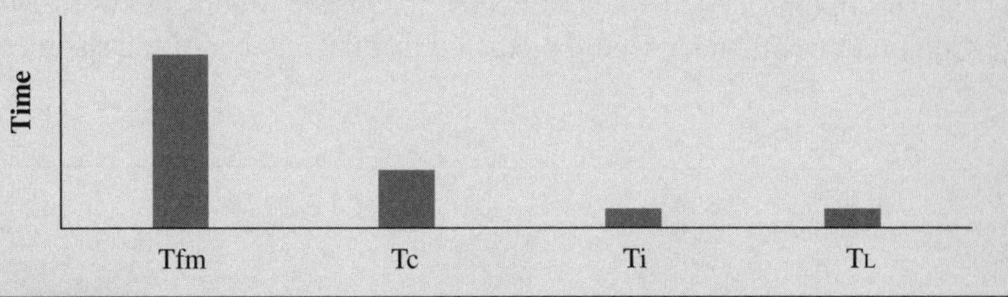

FIGURE 2-10 The computational times ranked by decreasing magnitude

The floating-point multiplication is clearly the most time-costly component. If, as in the weight example, we focus on only the single largest time item to increase computational speed, we would try to drive down the time required by the floating-point multiply operation. But notice, suppose that we were able to improve the floating-point multiply component so that it takes only 200 ns instead of 1200 ns—that is, make it six times faster. The time required to get through the loop is then 1000 ns, and we are not six times as fast but only twice as fast. As we reduce the large component, the other components become increasingly controlling and dominate. Even if we were to drive the floating-point multiply time down to 0, we would increase our speed by a factor of only 2½.

The moral of these two examples is that you should properly model and understand the best return that you can achieve in concentrating on one dimension for improvement. You don't want to over-promise and under-deliver to your sponsor (or yourself).

digital camera, a motor, and a battery. When you start your improvement invention activity, the relative costs of these four modules are as depicted in Figure 2-11.

The computer and the digital camera are big-ticket items as far as the component cost analysis is concerned. But instead of working to reduce the cost of either of these items, you should consider that many others out there also need a computer or a digital camera for their inventions or business needs, and they also want to see the cost reduced. That means that many entities are already working on both of those problems and the computer and the digital camera costs may be significantly reduced later if you allow others to do the work while you do your R&D. Less popular may be the motor and battery, and perhaps by the time you have brought those costs down, the computer and the digital camera costs may have fallen significantly without any effort on your part.

This optimistic outlook is embodied in Gordon Moore's Law. In 1965, Intel Corporation cofounder Moore predicted that the number of transistors on a chip

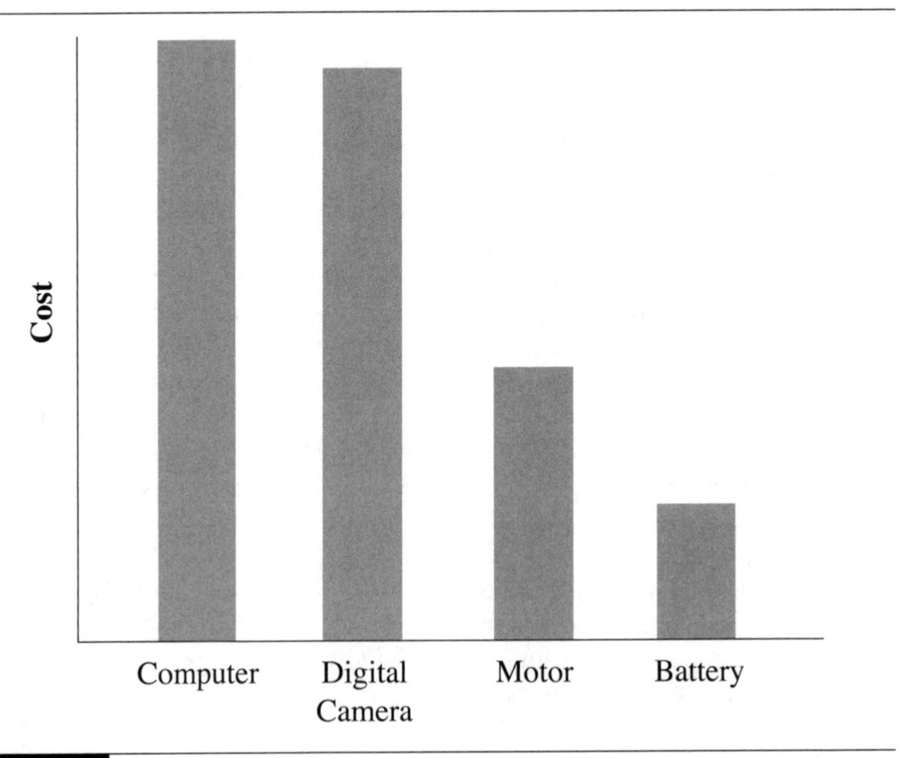

FIGURE 2-11 The relative costs of the four modules

would double every two years, a remarkable growth in transistor density. And his predictions came true.

From where do such remarkable advances come? To answer simply, from technology, is not an acceptable answer to the question. Technology brings up more questions than answers. To my thinking, at least, a new concept rallies technological enablers around it. The concept is a nucleus, a seed crystal in a supersaturated solution just waiting for crystallization. Consider Figure 2-12, for example, which plots the number of issued U.S. patents per filing year (1995–2000) that were concerned with digital cameras using a charge coupled device (CCD). Clearly there is intense, broad-based interest in continually perfecting the CCD camera, i.e., others are doing the work.

Tricks of IP Mining

One useful technique can significantly shorten your R&D time and your costs: *intellectual property (IP) mining*. If you can find an abandoned patent on something that is good enough for your application, or at least something that will jumpstart your R&D, you may significantly lessen your development costs and time to product.

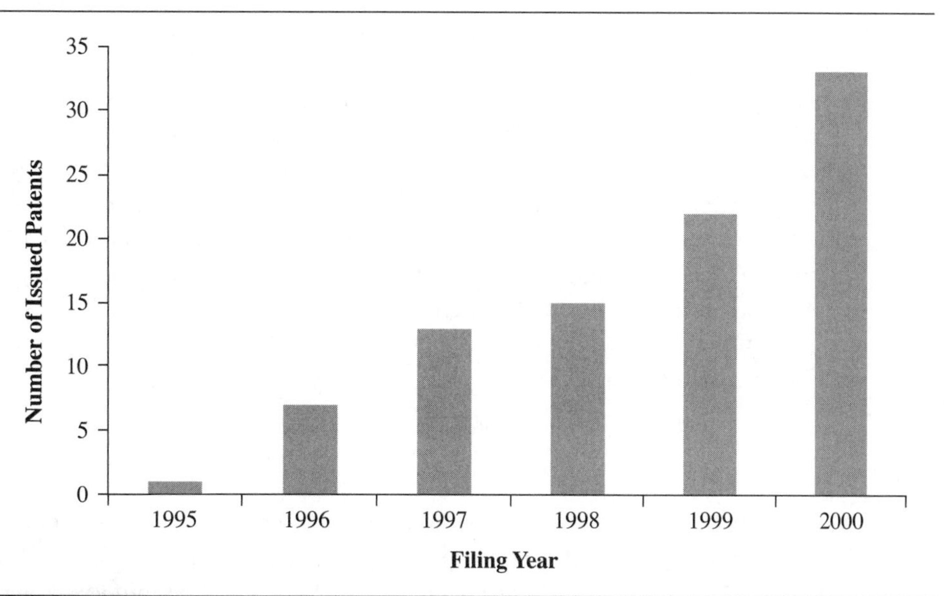

FIGURE 2-12 Number of CCD camera patents by filing year

The USPTO charges maintenance fees on utility patents, which are patents that teach the way in which an invention is used and works, and which are the bulk of patents applied for and issued. These maintenance fees are due at 3½, 7½, and 11½ years after the utility patent issues. If the fees are not paid in a timely way, the patent lapses and is said to become *abandoned*. At that point, the patent becomes incapable of infringement under abandonment, so *it belongs to the public*—and, remember, to gain a patent in the first place, the inventor had to teach someone of ordinary skill how to make and use the patented invention. This is your opportunity to "mine" this IP.

Let's return to the robot trash collector example. Suppose that the drive train engineers decide that the robotic trash collector motor should be a small, direct-current, electromotor. They do a search to winnow down candidate patents to those that have had a chance to become abandoned by searching up through 2005.

Their search returns U.S. Patent 6144133 "Direct Current Electromotor," as one (actually the only) candidate patent. The patent's content advertises that the patented device has a simple design and small dimensions. This sounds ideal, but do the engineers need to take a license to use this patent? To find out, they go to the USPTO's maintenance fee page (https://ramps.uspto.gov/eram/patentMaintFees.do). Following the instructions there, they found that the patent expired on 11/08/2004 for "failure to pay maintenance fees."

So it appears we might have come a long ways in reducing R&D costs and time, but there are two caveats with this technique. The first is the so-called "rocking chair patent" situation explained in Appendix A. Before we can produce devices taught by U.S. Patent 6144133, we must ensure that the patent was not "dominated by" another, broader patent that is still in force. The second caveat, although it is a highly unusual situation, concerns abandoned patent revival whereby an abandoned patent can sometimes be revived even after a significant period of abandonment. In such a case, however, there may be a question of what is known as "intervening rights" that can provide some shelter to this activity. Because this is a complicated legal area, it is best to proceed with analysis, advice, and guidance of counsel. But if you are successful at mining a patent, the cost and time savings can be huge.

Recap

We have studied improvement inventions and examined actual patents and some of the industries that depend upon improvement inventions. As long as technology keeps advancing, improvement inventions will be important, and this is demonstrated by the continual improvements of devices and methods used within and around large multifunction infrastructures. We also discussed how to focus our inventive abilities effectively on a specific dimension for improvement to make the most significant contribution.

Discussions and Reflections

- Survey the inventors in your workplace. Do they practice primarily on improvement inventions?
- What is the science or technology on which your business is based? What new single-dimensional technical advance would substantially alter it?
- Rank the problems in your ongoing work along a set of single dimensions. Where do you see the biggest gain for an investment of time or money? Carefully outline your strategy.
- I spoke of the ski business as an invention-persistent art. Can you think of other examples of invention-persistent arts?
- Consider an incandescent light bulb or any other relatively uncomplicated article. Make a list of as many dimensions of possible improvement that you can. You may be surprised at the length of your list. Have all of them been done? Should all of them be done? For example, U.S. Patent 3236699, "Tungsten-Rhenium Alloys," was issued February 22, 1966, on an application filed May 9, 1963. It is concerned with making an incandescent light bulb's filament "tougher." We usually associate tungsten with high temperature uses such as incandescent bulb filaments. We're less familiar with rhenium, element 75, which is a heavy silvery white metal. A rare element, it is extremely costly, typically pricing at thousands of dollars a pound.

  With alloys, a little of something can change something else a lot. So it is with rhenium and tungsten in the role of high temperature light bulb filaments. A filament, after all, has a tough job: It has to glow brightly and to do that it has to withstand a very high temperature produced by forcing an electric current through it, and that causes the filament to sag. Now, put this already stressed element under more dimensions of stress, such as shock or vibration, say as a light in a fan fixture, and you are asking a lot when you require it maintain its mechanical integrity. But a little bit of rhenium goes a long way toward diminishing the problem and this is reflected in the patent:

Useful sag-resistance distinguishes filaments of the invention from previously known tungsten-rhenium alloy bodies. The wire consisting of the alloy containing 3% rhenium is particularly useful for incandescent lamp filaments where, in the past, brittleness characteristic of all previous tungsten filaments used commercially for such lamps has been a major cause of failure of lamps by breakage of the filaments as a result of mechanical shock caused by vibration, impact, or both, incident to handling and shipment and during the service of the lamps.

- How do you assess and keep track of the disruptive technologies that could impact your improvement invention activities? Consider, for example, the Pony Express, which was not exactly an invention, but rather an innovation. The service was established in April 1860 with the eastern terminus at St. Joseph, Missouri, and the westernmost end at Sacramento, California. A year before the Pony Express started operations, St. Joseph had been reached by railroad and the town was then at the western edge of American civilization. The route from St. Joseph to Sacramento was nearly 2000 miles long, took a stagecoach 20 days, and the Pony Express service reduced this by one-half the time to transport a letter over the large distance.

  The service required enormous logistics with 165 way-stations, station-keepers, route superintendents, 400–500 horses, 183 riders, and their nourishment and support but the rapidly advancing telegraph technology soon made such a delivery mode moot. After it was over, the business had cost about $300K and returned less than $91K. It was discontinued in October 1861, merely an 18-month affair. The telegraph was the disruptive technology that killed the Pony Express.

Chapter 3

Developing an Inventive Mindset by Gaming the System

In Italy, for 30 years under the Borgias they had warfare, terror, murder, bloodshed. They produced Michelangelo, Leonardo da Vinci, and the Renaissance. In Switzerland they had brotherly love, 500 years of democracy and peace, and what did that produce? The cuckoo clock.

—*The Third Man*, 1949

Some say that every successful invention is accomplished by fulfilling a perceived need. This need can be generated in the traditional sense, in which a large portion of society acts as a collective catalyst to urge the invention (see Figure 3-1). But inventions can come about in other ways. Sometimes inventions result from inspirations generated by the desire to "game the system" and by the complementary desire, or need, to prevent the system from being gamed.

What do I mean by "gaming the system"? The *system* is a collection of established methods and behaviors and a set of rules that govern people, tools, and processes. Rules are the modes and mores, the traditional behaviors that establish how the *game* is played. Most people follow the rules of commerce and custom by faithfully staying within the bounds. But sometimes someone will realize that the system can be exploited, or gamed, for personal gain.

Of course, this can result in illegal activity or chaos, but gaming the system can also lead to innovation, and new inventions can result from such behavior. Often a

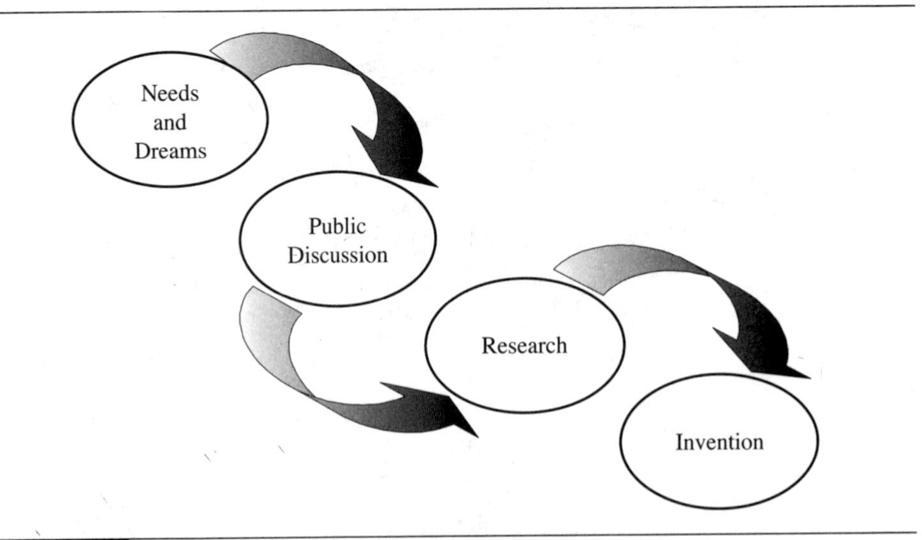

FIGURE 3-1 Successful inventions fill perceived needs

backlash of sorts follows successful system gaming, with inventions or innovations that try to prohibit or otherwise discourage disruption to the established norm. This cycle of gaming the system and preventing the gaming has existed in many forms throughout history. In the military, it might be categorized as "countermeaures" and "counter-countermeasures." In taxation, it is the ever-growing corpus of tax laws and the art of tax loopholes.

Once you thoroughly understand a system, you might have a Eureka! moment that can lead to your gaming the system to introduce your innovation. This principle applies equally well to a legitimate entrepreneur as it does to a scam artist. In one way or another, gaming the system is the way we do business.

I have had success in generating Eureka! moments by first understanding a system and then searching for a way to game it. I suspect that every successful inventor has a bit of a rebel nature. After all, we produce inventions that are not obvious to most people, and any work product that is not obvious, however slight the degree, is an inherent challenge to the well-ordered plans and timetables of most managers. This activity is *creative thought*, and it's particularly important for human progress; nevertheless, it can't be budgeted, expected, or well measured by any of the de-dimensionalizing metrics such as dollars or man-months that are so much in vogue among the business elite.

Creativity often seems to thrive best in an agar of chaos, which bothers the financial watchdogs to such an extent that in many industries, innovation is controlled and must be undertaken in such a way as to be devoid of any components that cannot be measured, sequenced, and controlled. Creative thought, however, is not an immediately productivity-enhancing activity, which poses another problem for corporate bean counters. It is unsettling to reflect on how many geniuses in history must have died behind the plow, and how many great seeds of ideas were never germinated.

In this chapter, we will look at a number of examples in which the system was gamed. My hope is that this will help sensitize your thinking about this mode of innovation. We will study an invention produced by an inventor using this mindset, and we will discuss existing problems that are awaiting your Eureka! moments to help prevent the gaming of an important system.

Examples of Gaming the System

The following examples demonstrate opportunities for Eureka! moments. In analogy with the military mode of countermeasures and counter-countermeasures, the first opportunity is to innovate so as to game the system. The second opportunity is to innovate a way to prevent or counter the gaming of the system. The examples show how gaming the system can be prevented by changing the rules of the game—in this case, imposing new regulations, which is also a type of innovation.

Gaming the Game

Sports and games provide perfect examples of established systems that can be "gamed." By exploiting the rules and regulations of the gaming system, you can create an innovation. We'll examine the games of baseball and golf to see how best to exploit these systems to create innovation.

Baseball's Infield Fly Rule

The history of the great American pastime, baseball, provides an excellent example of gaming the system, or, more accurately, gaming the game. In the late 1800s, a baseball game had gotten to the point where there was one out and two men on base, one on first, the other on second. The batter popped a fly and the infielder positioned himself to catch the ball. The men on base were anxious. At that time, the dimensions of the baseball playing rules did not preclude a fielder's intentionally failing to catch the ball. Should the men on base run or should they hold? If the infielder caught the fly, as he certainly could, the batter would be out and they had better be on their original bases or the side could be retired by two outs gained by a throw to first or second. But if the infielder were to intentionally drop the ball, as he certainly could, then the side could be retired with two outs by a forced out at third and second.

The infielder could therefore "game the game" and create two outs instead of just a single out. It was a true gaming of the system, a leveraging of the rules that the leagues couldn't abide; and in 1895 the game was modified to include the infield fly rule and take away this dimension of play by declaring the batter out under these circumstances.

Golf Tees

The second example has to do with an actual patent—specifically, U.S. Patent 1599207, "Golfing Tee," issued September 7, 1926, from an application filed March 26, 1925. The patent is for a golf tee that is designed to lengthen the golfer's drive.

Figure 3-2 shows a figure from the patent application. The numbered elements correspond to the following:

1. The tee's shank
2. The tee's point

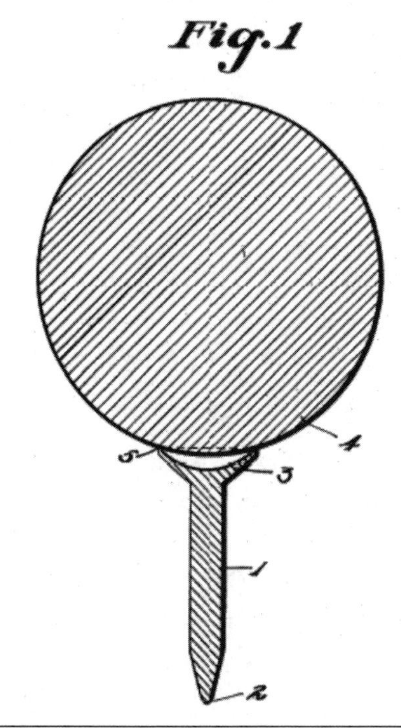

FIGURE 3-2 Figure 1 of U.S. Patent 1599207

3. A cup-shaped top
4. The golf ball supported by the improved tee
5. A sharp rim of the improved tee

The prior art comprised sand tees that conformed to the golf ball's surface and allowed the ball simply to slide off the tee when hit without inducing rotation. The inventor, Samuel Capen, motivated his invention, an improved golf tee, with the following:

In my improved golf tee, when the ball is driven, the tendency is to roll over the forward edge of the ball supporting rim. This creates what is known as top spin in the ball during its flight. Top spin or rotation in the direction of travel is a highly desirable condition because its effect is to increase the distance that the ball will roll after it has struck the ground, thus lengthening the drive.

Capen's Eureka! moment came when he realized that adding a sharper edge at the rim (number 5) would resist the ball's motion off the tee when the ball is hit, and this resistance would induce the rotation by forcing the top of the ball to travel faster than the bottom of the ball. The new tee seemed to provide an advantage to the golfer, but as with the infield fly rule, the rules of the game caught up: According to Golf-information info "The development of the tee was the last major change to the rules of golf" and the rules of golf now disallow the new tee as:

> *A "tee" is a device designed to raise the ball off the ground. It must not be longer than 4 inches (101.6 mm) and it must not be designed or manufactured in such a way that it could indicate the line of play or* influence the movement of the ball.

As you shall see in the following examples, prevention of the gaming of the system is not always quite so straightforward.

How Criminals Game the System

Friedrich Nietzsche wrote, "The criminal is quite frequently not equal to his deed: he belittles and slanders it." But when the criminal is equal to his deed, watch out! Most criminals are not very clever. But when someone is clever and willing to game the system by stepping across the line to illegal conduct, he or she may pose a threat that can be quite formidable.

By presenting these examples, I'm not suggesting that you use your talents as an inventor to become a master criminal! In fact, your inventiveness, your Eureka! moments, are desperately needed to counter crime, which is becoming ever more sophisticated with an ever greater deleterious economic impact. My point here is that if anyone exemplifies the mindset of inventiveness in order to game a system, it is the criminal.

For example, in 1888, an Inspector Bonfield told a *Chicago Herald* reporter that "It is a well known fact that no other section of the population avail themselves more readily and speedily of the latest triumphs of science than the criminal class. The educated criminal skims the cream from every new invention, if he can make use of it."

Bonfield's admonition is as valid today as it was more than a century ago. Consider, as of a few years ago, U.S. crime statistics reported that if a robber went into a bank with a gun and robbed it, he could expect, on the average, to make off with about $5000. Of course, he might get more, or he might get less; and he might get shot. And if he's caught, he goes to jail.

However, if that same person went to school and studied accounting, bought a suit, and got a job as an accountant and practiced not double-entry bookkeeping, but rather double bookkeeping (aka "embezzlement"), then he could expect to make off with about $19,000. This is almost four times the amount that he'd get using the more antisocial armed approach—plus, if he got caught, he'd probably go to a nicer jail!

On the other hand, if this person got the proper education and was really good, he could try computer crime. At last check, a personal identification number (PIN) for manipulating funds had a pretty high "street" price. He might make $100,000 using that PIN. And if he got caught? Well, the authorities might be willing to make him a consultant, as they did with Willie Sutton, the infamous bank robber.

The following two examples represent the tip of the iceberg in a mode of crime referred to as the "man-in-the-middle" syndrome. This particular crime gambit is an illegal gaming of the system that is difficult to counter. It threatens most of us, especially as we turn our paper dollars into bits and bytes and make our deals by way of electronic transactions.

The Man in the Closet

Many years ago, before credit cards, large retailers were limited to accepting checks or cash only. Some old-time stores used a most interesting system: a customer would purchase an item on the top floor, the payment and the paperwork would be placed in a pneumatic tube and sent to the basement, where a finance clerk would record the transaction, make change, and return the change and receipt also via the pneumatic tube to the sales clerk on the top floor. The system was extremely efficient and reasonable for the technology of the times.

One store had an usual setup, however. The pneumatic tubes passed through another floor and, in fact, behind a janitor's broom closet on that floor.

The janitor at this store discovered that he had a most special broom closet when he realized that behind the wall were pneumatic tubes through which coursed the lifeblood of the store. Well, every so often, our friend would simply interrupt the down tube, throw away the paperwork, take the correct change out of his own pocket, and send it back up in the other tube (as shown in Figure 3-3). The unsuspecting sales clerk was happy. The customer was happy. And the finance clerk was happy, because what she didn't see, she didn't log.

The store lost about $100,000 before it discovered and shut down the broom closet operation. Back then, $100,000 was a huge amount of money. So why did it take so long to discover these losses? The answer is found in the dimensions of its business model. The losses simply got lumped into that great overhead catch-

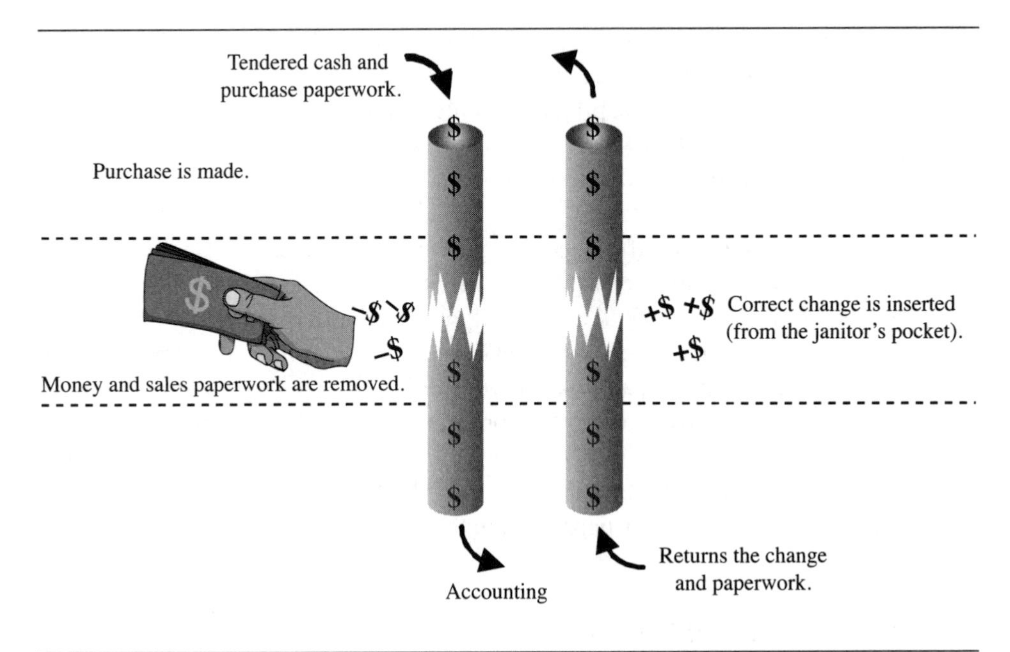
Tendered cash and
purchase paperwork.
Purchase is made.
-$ -$
-$
Money and sales paperwork are removed.
+$ +$
+$
Correct change is inserted
(from the janitor's pocket).
Accounting
Returns the change
and paperwork.

FIGURE 3-3 The department store with the man-in-the-middle

all dimension: shoplifting. The game ran because of the elasticity of this slack variable in the system.

In the old department store, cash flowed as paper and coin. Many years later, but still many years ago, cash started flowing digitally and the genie was out of the bottle.

Cryptographic Security and International Banking

Security is just plain counterculture to the American psyche. Adding security is observable by the mainstream only in that it adds cost and some inconvenience—and neither of these is in great demand. You cannot really measure or quantify the advantages of security. Everyone knows that it is a good thing because it makes us less vulnerable with it than without it. But less vulnerable to what? And if I were less vulnerable, would I be a viable target? If a competent opposition were exploiting me, would I even know it?

Most folks, when they think of security matters, think of spies and defense plans. Although that certainly is part of the history, times change, and if you were a competitor nation, would you rather have (a) the plans to some new missile that probably won't be built, and even if it were built, it would probably not function anywhere near its design envelope, or (b) an insight to the thinking and plans of the Federal Reserve Board? Economic intelligence is probably worth its weight in gold—or at least in oil.

The following case is an example of gaming a financial system that had supposedly been hardened by a countermeasure—a cryptographic authentication scheme.

A number of years ago, I received an invitation from a vice president of a major bank in New York City to attend a meeting on electronic funds transfer security. I was invited to make a presentation to representatives of several large banks about a type of cryptography that they thought might be useful.

After my presentation, I spoke privately with the fellow who had invited me. I told him that one of the frustrations of working in the security business was that although you postulated a threat, you often did not know if the bogeymen really existed: If "they" are truly competent, then, of course, you wouldn't know what "they" were up to.

My host was quite open and said that indeed he had met the threat in form and function and it had a real interest in subverting dollars out of the ether. He went on to recount an incident that occurred in one of the large electronic funds transfer networks. This particular network used cryptography for authenticating messages. Cryptography is that very old art turned science that deals with two things: a process that renders messages unreadable except to authorized parties, and the authentication of messages, to validate who the sender claims to be with all of the implied rights and privileges.

Unfortunately, the system had adopted a poor cryptographic authentication scheme. Just how bad became evident one day when a message arrived in a major U.S. hub by satellite relay from a bank in a foreign capitol. The message was addressed to an action bank, and to reach its destination the message had to pass through a number of banks in between. At each of these intermediate banks, the message authentication was independently checked. The scenario is pictured in Figure 3-4.

The message was received at a clearinghouse and relayed on to the action bank. The message directed the action bank to transfer several million dollars from one account to another. The message's authentication was checked electronically at each intermediate bank in the relay, but it ran into an interesting problem when it hit the action bank. This particular bank had not yet installed its electronic checking apparatus, and the bank personnel were handling messages the old-fashioned way: They looked at them.

One of these folks just couldn't reconcile himself to the validity of such a transfer, even though the cryptography seemingly proved it was the duly certified

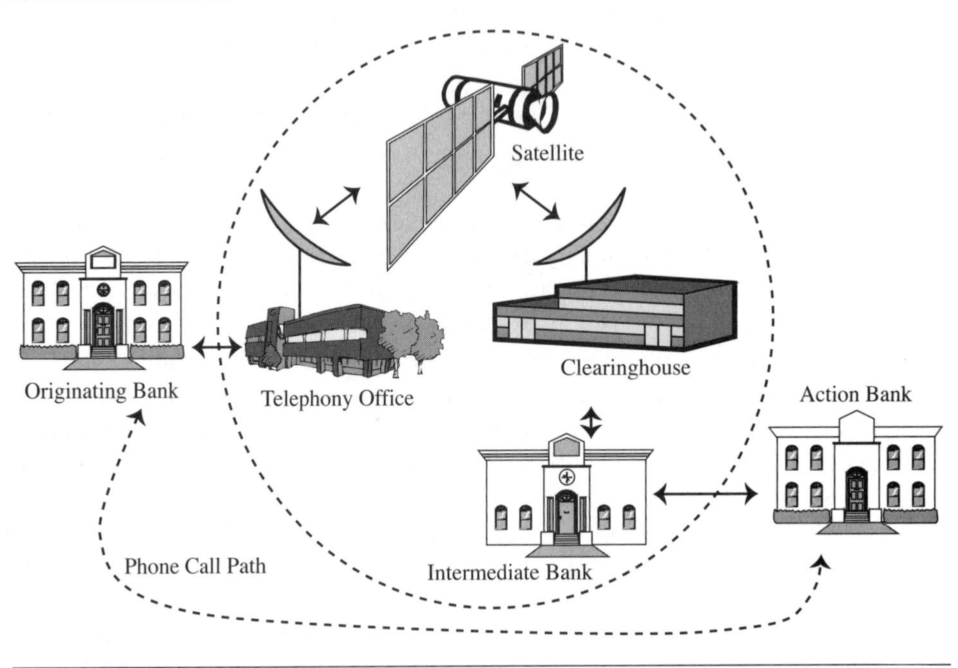

FIGURE 3-4 The path of the messages

intention of the issuing bank. So the action bank drafted a cable to the issuing bank, to the effect of "Do you really mean this?" The answer came back a few hours later, again properly authenticated, acknowledging that, yes, it was indeed correct, and, while you're at it, add a few million more to the transfer. (Some days it just doesn't pay to ask.)

Our friend still couldn't accept it and placed a phone call to a colleague he knew at the originating bank. And, surprise, surprise, the originating bank hadn't originated any such order. At this juncture, the investigators were able to draw a line encompassing every stop between the two banks. Somewhere in the compass of that boundary lay the problem. (They probably exempted the satellite, as it's cold and lonely up there.)

The answer turned up and was interesting. A couple of folks at the foreign country's telephony office had gone into a sideline occupation and developed a game that was undergirded by the assumed trust in the system's impotent countermeasure. They had studied the authentication process and determined it was not very sound—indeed, it was eminently exploitable, and they proceeded to exploit it. As the old joke goes, they probably went from a nine-to-five job to five-to-fifteen.

Back in 1980, $19 trillion was passed by paper check in the United States, but in that same year, $117 trillion passed electronically. To accommodate all of these transactions, and the others that are ancillary or independent, requires a lot of circuits, or "tubes" as our friend the janitor might call them. I suspect that there are also a lot of "broom closets" with lots of room and electronic opportunities for the modern "man-in-the-middle."

How to Seek a Eureka! Moment by Gaming the System

The human being is a complex system, but like most other complex systems, humans can be gamed. To come to a Eureka! moment, you need to understand the system, the human being in this case, and to find that dimension of the system that can be gamed. One very popular and lucrative gaming of the human being is through gambling and betting. Why is gambling considered gaming of the human system? I believe it falls into capitalizing on the human desire to believe in luck—the idea, concealed in hope, that the spin of a wheel will be favorable, that cards dealt will be winners, or that a scratched lottery ticket will reveal instant fortune.

A group of us centered an invention quest on devising a game that would meet two goals: the game must appeal to the human being's notion that he or she can be successful in playing the game, and the game must be profitable for the house.

We first sought to answer the question: What do gamblers believe can be used to their advantage? We settled on an appeal to a basic belief held by competitors that the first party to make a move, the first-mover, generally has an advantage. This belief is not without empirical foundation, and two ancient board games, chess and Go, both seem to obey this notion.

In chess, white is the first-mover. Many players and theorists believe that the player beginning the game with white has an advantage. Statistics gathered over the last century show white beating black 52 to 56 percent of the time.

In Go, the first-mover is black. According to Rob Van Zeijst (www.yomiuri.co.jp/dy/columns/0001/415.htm):

> *In most games, including chess and checkers, the person who moves first has an advantage. This also used to be true for Go through most of the game's 3,000–4,000-year existence. However, a system was devised to make up for Black's first-move advantage by giving White some points as compensation. These points are called komi. To prevent draws, Black gave 4.5 komi in the beginning. These komi were increased to 5.5, but 6.5 is more common nowadays. Some*

tournaments have an even larger komi, while others allow a player to choose what color he wants to play. The official change in komi to 6.5 was the result of the shared opinion among the strongest players in the world.

We sought to invent a game in which the player would make the first move (and think he or she had the advantage), but no matter what move the player made, the house would actually always have the advantage. We based our search on some esoteric mathematics involving nontransitive properties of sequences and we were successful. Our Eureka! moment led to an invention that is a form of gaming of the human system.

NOTE *For those interested in the details of the mathematics, they are published as "How Many Random Digits Are Required Until Given Sequences Are Obtained," by G. Blom and D. Thorburn (Journal of Applied Probability, Vol. 19, pp. 518–531, 1982) and reviewed in Data Transportation and Protection, by John E. Hershey and R.K. Rao Yarlagadda (Plenum Press, 1986).*

The setting chosen for the new game invention is couched as a derivative game for a casino that has a roulette wheel. Our game is not predicated on the game of roulette, the roulette wheel is to function as normal and drive the action at the roulette table. We use the wheel only derivatively to provide a random series of RED and BLACK outcomes, each with a probability of 50 percent. (We ignore GREEN outcomes from the roulette wheel.) The stream of randomly and fairly generated RED and BLACK outcomes is transmitted from the roulette wheel operations to remote units that are playing our game as illustrated in Figure 3-5. An observer can view the roulette wheel and see that its random outcomes are not altered but transmitted as actual outcomes. Also, from an engineering perspective, many modern roulette table operations have an infrared sensor that automatically reads the roulette wheel outcome and electronically displays each outcome at the table.

Each remote game unit has a display module as shown in Figure 3-6. The player goes first and enters three RED (R) or BLACK (B) markers, P1, P2, and P3 into the game unit's display window. The markers may be all black, all red, or a mixture of black and red. The game module goes second and selects three RED or BLACK markers, G1, G2, and G3, and enters them into the game unit's display window. Similarly, the game module's markers may be all black, all red, or a mixture of black and red.

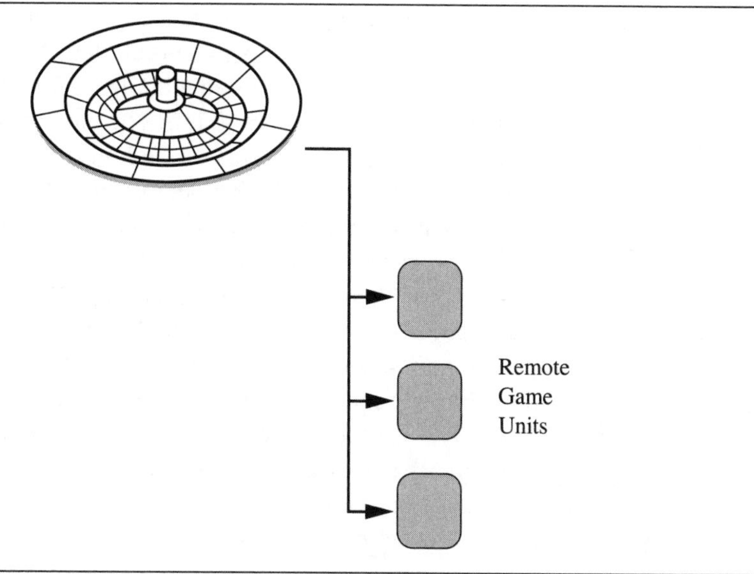

FIGURE 3-5 A roulette wheel providing RED and BLACK outcomes to the new game units

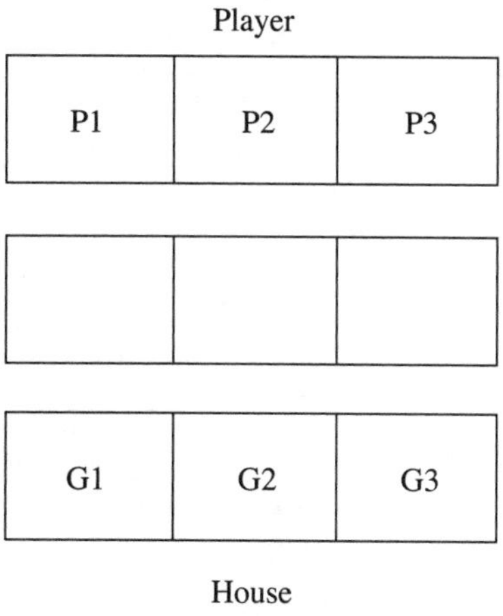

Player

P1	P2	P3
G1	G2	G3

House

FIGURE 3-6 Display module of the new game unit

After both the player and the game module have entered their markers into the cells of their respective windows, the game is started. The RED and BLACK symbols from the roulette wheel are now sequentially shifted through the middle window of the display module from right to left. The game is won by the player or the house whenever the player's markers or the house's markers match the symbols in the middle window. Let's look at an example. Suppose that the player selected and entered RBB into the player cells of the display module as shown in the top-left drawing of Figure 3-7. Proceeding left to right, top to bottom of Figure 3-7, suppose that the house then responded by entering RRB into its cells of the display module. The game is now started and the RED and BLACK symbols from the roulette wheel are now shifted, right to left, through the middle window of the display module. For our example, we will assume that the successive roulette wheel symbols are B R R B. You can follow the progression of these symbols in the drawings of Figure 3-7 and you will see that the game is terminated with a house win in the bottom-right drawing as the symbols in the middle window match the symbols initially selected by the house.

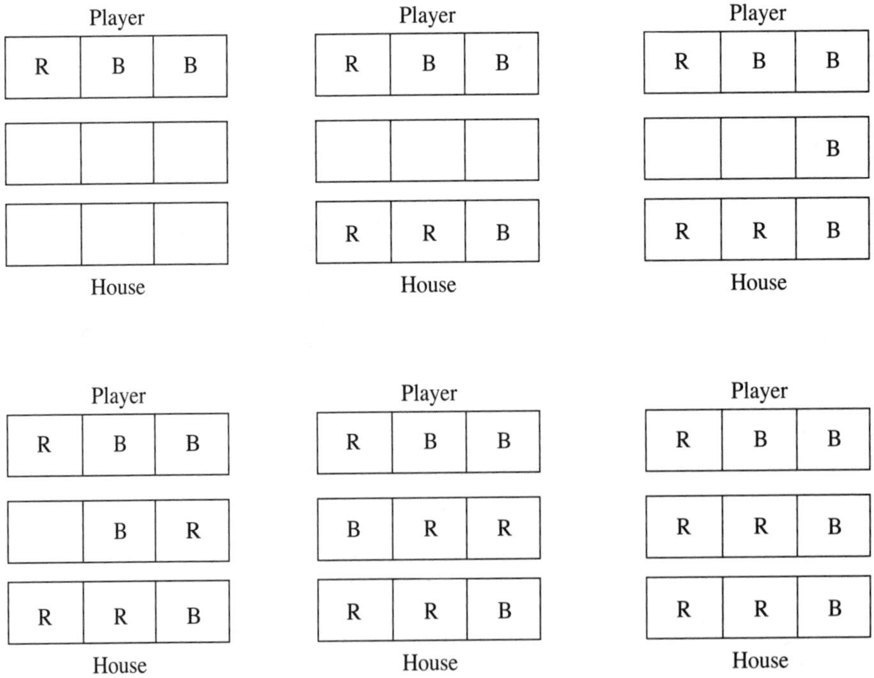

Player

R	B	B

House

Player

R	B	B
R	R	B

House

Player

R	B	B
		B
R	R	B

House

Player

R	B	B
	B	R
R	R	B

House

Player

R	B	B
B	R	R
R	R	B

House

Player

R	B	B
R	R	B
R	R	B

House

FIGURE 3-7 An example progression of the game

And now for the heart of the invention: The game module selects the house's marker symbols based on the first-mover's (player's) marker symbol selection according to the table shown in Table 3-1.

By using Table 3-1, the house will bet on marker symbols RRB because the first-mover player bet on marker symbols RBB. Further, the house expects to win two out of three times if the player bets RBB. By examining the Table 3-1 you will see that the house will always have an advantage over any first-mover bet if the house bets the marker symbol selection indicated to bet against the player's selection. This is a counterintuitive result, i.e., whatever three-marker symbol pattern the player picks for the player's pattern, the house will always select a marker symbol pattern such that the house has a greater than 50 percent chance of winning against the player. The reader is invited to simulate the game to be convinced that this is so.

Recap

We discussed a concept I call "gaming the system." You can think of inventions as ways to game a system—finding a way that conventional rules or usual operations may be side-stepped, modified, or even turned completely on their heads in order to achieve an advantage. Sometimes this produces an invention of value, and sometimes it opens a hole that must be covered by perhaps another invention. After you thoroughly understand a system, you will understand its prejudices and shortcomings and the ways in which it may be improved, hardened, or expanded with gaming to reach your Eureka! moment.

Player Symbol Selection	House Symbol Selection	Probability House Wins
BBB	RRB	7/10
BBR	RBB	3/4
BRB	BBR	2/3
BRR	BBR	2/3
RBB	RRB	2/3
RBR	RRB	2/3
RRB	BRR	3/4
RRR	BBR	7/10

TABLE 3-1 House Marker Symbol Selection

Discussions and Reflections

- The law and molecules—gaming the law: Q. What changes faster than the law? A. Almost everything else.

  Consider the Controlled Substances Act, which became U.S. law in 1970. In 1971 the molecule for amphetamine was listed as a controlled substance. But by making a small change in the molecule's structure, specifically making a ketone transformation by adding a single oxygen atom, as shown in Figure 3-8, we arrive at cathinone, also a psychostimulant that is used recreationally. But cathinone was not placed under the Controlled Substances Act until about 1993.

  The problem is that it is generally easy to produce a molecule very close in structure to a controlled substance that enables recreational remand but is not listed as a controlled substance and therefore legal, simply because it is not illegal. The substance might be extremely hazardous to your health and general well-being, but if you insist on using it recreationally without FDA or other health authority vetting, you can do yourself serious physical damage.

  The business of producing legal chemical highs is a thriving enterprise and, unfortunately, an excellent example of gaming the system where the system is the corpus of drug laws. Can you think of any similar examples where a system is constantly gamed?

- Consider the invention-inspiring squirrel versus the lowly bird feeder. Many feeders are patented, including a number of design patents. Bird feeder invention falls in what is termed a "crowded art." In 2006, I made

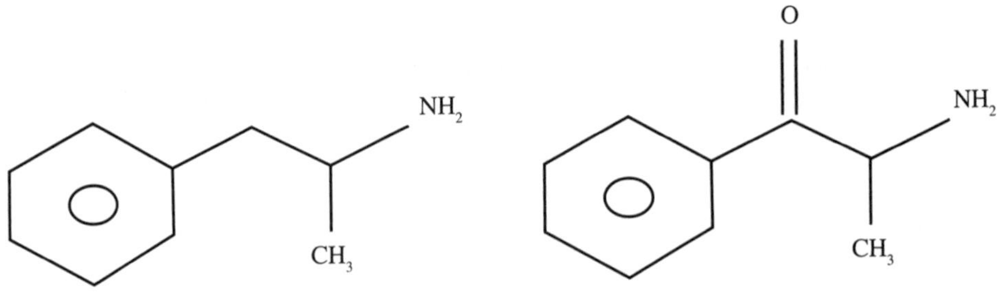

NH_2

CH_3

Amphetamine

O

NH_2

CH_3

Cathinone

FIGURE 3-8 Synthesizing cathinone from amphetamine

Filing Years	"Bird feeder"	"Bird feeder" and "squirrel"
1990–94	133	19
1995–99	127	21

TABLE 3-2 U.S. Patents Regarding Bird Feeders

some counts of the number of issued U.S. patents that had the phrase "bird feeder" in their claims, title, or abstract. And of those, I counted the number also specifying the word "squirrel." I found the data shown in Table 3-2.

Why do we find such a strong coupling between bird feeder inventions and squirrels? I believe this is an incidental societal stasis. People are enamored of birds, and squirrels continue to game the bird feeders and thereby bring out the inventiveness of people as they continually seek to "game the squirrels." See if you can find other examples of persistent and unquenchable motivators for invention.

A Growing Concern: Flowerpots

Someone might object that filing 133 patents on bird feeders in the first five years of the 1990s might exhaust the possible white space and severely reduce the number of inventions possible in the second half of the decade. *Au contraire!* The remarkable thing about white space is that it keeps rolling before us, an endless plain awaiting the plow of invention. Donald Weder has demonstrated this in spades, or perhaps I should say flowerpots. Weder is perhaps the most prolific of American inventors if judged by the number of issued U.S. patents. He has been issued more than 1300 patents, and, remarkably, most of them are for flowerpots or floral groupings. He filed 360 of the issued patents in the years 1990–94 and 535 of them in the interval 1995–99.

- How do you find logical and out-of-the-box thinkers who are good at inventing by gaming the system, especially when you interview a candidate for only a half an hour? Try this exercise on the applicant:

  Look at the drawing in Figure 3-9. There are two rooms. A person is initially in the room on the right, where there are four switches connected to four incandescent lamps in the room on the left. An opaque door lies between the rooms and is closed. The person is twice allowed to manipulate

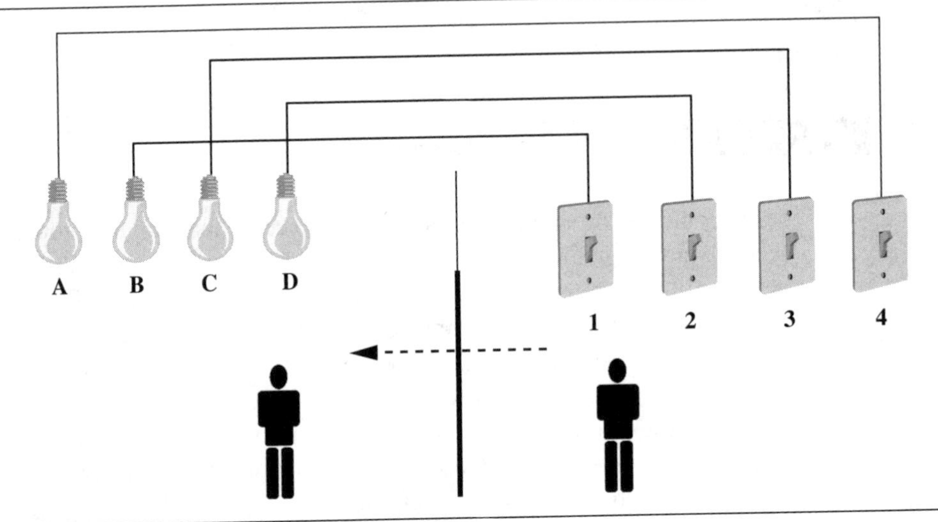

FIGURE 3-9 The light switch connection problem

the switches and enter the room on the left and inspect the lamps. The door is closed between visits. How does the person discern which lamp is connected to which switch?

This is a great interview question because it tests a person's sequential logic skills. There are a number of possible answers. The person could initially set switches 1 and 2 to the "on" position and enter the room. The person would observe that lamps B and D were lit. The person could then return to the room on the right and switch 2 to the "off" position and set switch 3 to the "on" position. On re-entering the room on the left, the person would now observe that lamps B and C were lit. As lamp B has remained lit, it must be connected to switch 1. As lamp D had been extinguished it must be connected to switch 2. As lamp C had become illuminated, it must be connected to 3 and that leaves switch 4, which, by the curiously named *pigeonhole* principle must be connected to lamp A. Bravo!

Now for the out-of-the-box follow-up question. Same scenario, same rules, except that this time only three switches and three lamps are used, and the person is allowed only one visit to inspect the lamps in the room on the left. What lamp is connected to what switch?

Turns out that the real logical types will fail the second part. They will take the problem to a logical construct in their mind, mapping the words of the problem to a limitation of their scope. There are two key words in the problem that carry extreme weight. The words are "incandescent" and "inspect." Why were they chosen, what do they ensure, and what do they allow?

First, the bulbs are *incandescent* lamps. This means that they derive their light output from electrically heating a filament—that is, if they are left on, they get warm. Second, the person may *inspect* the lamps, not merely *look at* them. So, all the person has to do is to set switches 1 and 2 to the "on" position, wait a bit, turn switch 1 to the "off" position, and enter the room on the left. By *looking* to see which lamp is lit, the person knows that that lamp is connected to switch 2. By then *inspecting*, by touching the two dark lamps, the person knows that the warm unlit lamp is connected to switch 1 and the colder lamp to switch 3.

The important lesson here is the dimensionality of the words of the language. The word "inspect" has a much greater dimensionality than the word "look." When inventing, carefully parse the language of any restrictions or limitations to see if there is a way to game the situation. Also, when writing patent applications, you should be sure that the breadth of the words that you use is the breadth that you need—not too narrow, and not too far-reaching.

Chapter 4

Increasing Dimensions to Spark Eureka! Moments

We thought we understood two *because we understood* one *and* "one *and* one *are* two." *What we are finding is that we must learn a lot more about* and.

—Sir Arthur Eddington, Astrophysicist

As you'll recall from Chapter 2, the term "dimension" is not limited to the usual considerations of height, length, width, and perhaps time. I use the word in a very general way. As applied to the discipline of invention, dimension includes any and all distinct major components or features of a device or method, such as weight, power consumption, speed, basic technology, or just about anything else that is a feature of the device or method. As we consider combination inventions in Chapter 5, you'll see examples of combining dimensions of one device or process with another to create a combination invention that offers more dimensions than the composing parts.

In this chapter we will get acquainted with spotting different types of dimensions and the value that their inclusion can incorporate. We will also explore the question of properly selecting or changing a dimension in order to nurture a Eureka! moment.

Examining New Dimensions

Although they are not themselves inventions, the following three examples illustrate the value that can result from increasing dimensionality or, conversely, the loss in value that can result by using too few dimensions, which I call "dedimensionalization."

Language Dimensions

The purpose of language is to enable the transfer of information between people. Over time and with usage the building blocks of our spoken language, our words, may become imprecise. The imprecision is often a result of dedimensionalization resulting in ambiguity. For example, how do you speak about the weather? "It sure rains a lot in Seattle." But does it? What does "rain" mean? Consider the following: The mean precipitation for a year is 38.44 inches in Seattle. But the mean yearly precipitation in New York City is 44.22 inches. Is Seattle getting a bum rap? How did Seattle get this reputation? The statistics don't seem to bear it out.

But what statistics? What do we think we are comparing? What is it we really want to compare? Can a comparison in only one dimension be meaningfully made? The graph in Figure 4-1 shows a two-dimensional analysis of the annual rainfall in some major U.S. cities. Along the vertical axis is the mean number of

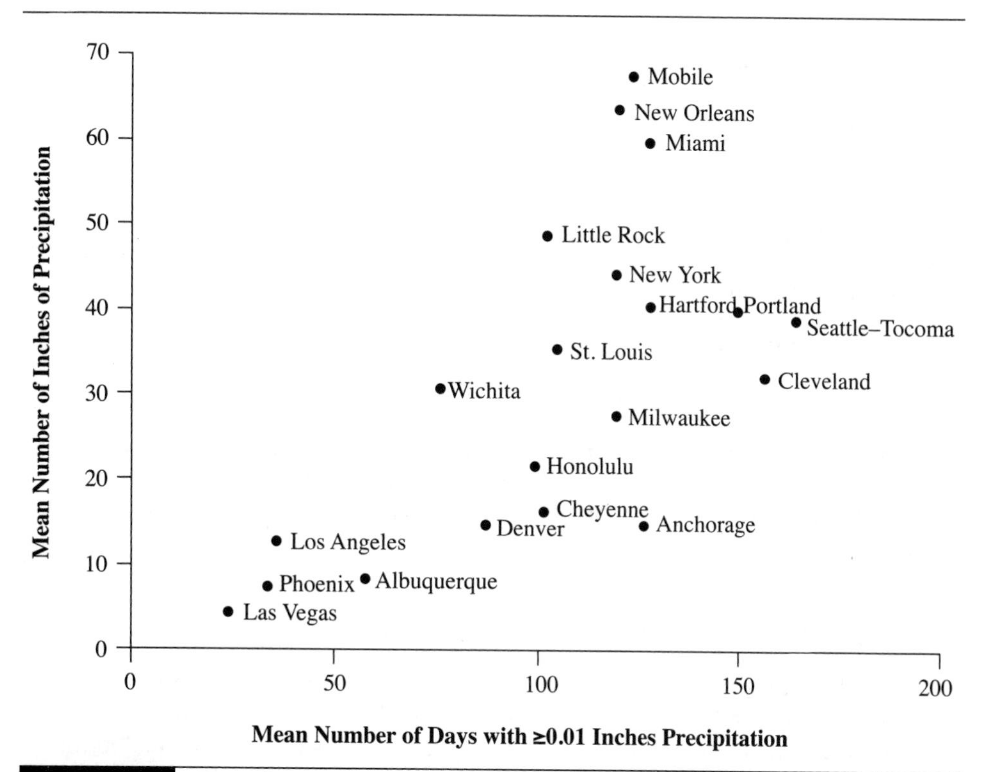

FIGURE 4-1 Two-dimensional analysis of rainfall

inches of yearly precipitation. Along the horizontal axis is the mean number of days on which there is at least 0.01 inch of precipitation.

Notice the mischief that can be made when considering more than one dimension. For example, if we compare Hartford, Little Rock, New York, Portland, and Seattle-Tacoma on the basis of the mean number of days with at least 0.01 inch of precipitation, we find that the order is as follows:

1. Seattle-Tacoma
2. Portland
3. Hartford
4. New York
5. Little Rock

But if we order in terms of mean yearly inches of precipitation, the order is exactly reversed:

1. Little Rock
2. New York
3. Hartford
4. Portland
5. Seattle-Tacoma

The Eureka! moment that comes from this example in speaking about the weather, is that more than one dimension must be considered in order to precisely use a simple and common word. There are many dimensions and inherent ambiguities to our speech. Did you ever wonder why legal fine print is so long and tedious?

Sales Savvy

We just looked at words and the problems that can arise from dedimensionalization. Similar problems can affect our paths in life and business. We may be driven by a goal and not realize that its successful pursuit requires us to proceed in more than one dimension. I learned such a lesson many years ago when I worked in sales for a fast-growing company.

I was part of the sales staff of a high-tech company that offered technical advice and solutions to companies large and small. As part of my research, I identified some technical problems I believed were commonly experienced by companies and sought to sell them our solutions. At this point in time, I was experiencing a sales slump, and I could not identify what was missing in the sales equation. I hadn't been able to turn a contract for an uncomfortably long time. I knew some of my target companies needed our help, and I knew we could offer them what they needed. What was missing?

I discussed this problem with one of our sharpest senior executives during a long international flight. I vividly recall his reply. With just a few strokes of his pen, he unveiled a simple and powerful two-dimensional tableaux, which is shown in Figure 4-2. He placed a large check mark in the upper-left box, and with this simple sketch, everything became clear to me.

I had been too close to the problem, and I had been considering only one dimension at a time, never both dimensions simultaneously. I believed in the

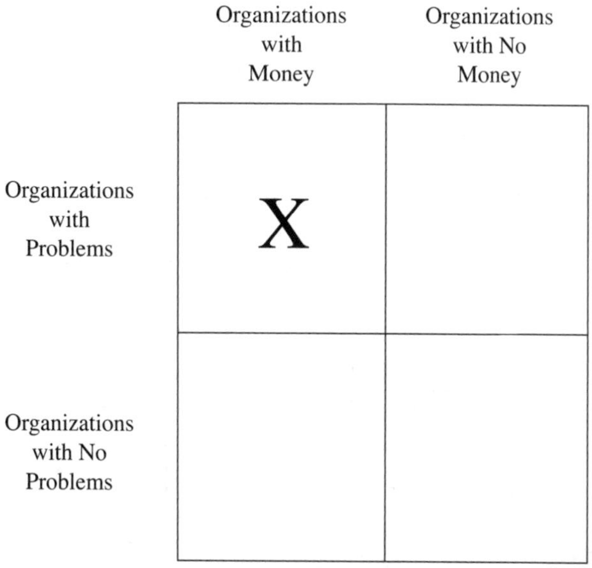

	Organizations with Money	Organizations with No Money
Organizations with Problems	X	
Organizations with No Problems		

FIGURE 4-2 Two dimensions for marketing

products and services that my company offered, and I was determined to sell these solutions, but I had been focusing on the quality of the products instead of focusing on the target companies' needs and purchasing power. I could not sell products to companies without problems or companies without the bucks required to purchase our solutions. After this Eureka! moment, I realized that I needed to refocus my efforts; I needed to see my potential customers in their proper dimensions.

A Combinatorial Conundrum

The insight and power that can accrue from adding the right dimension to your analysis or concept is immense. The right dimension can often provide a coign of vantage that renders obvious what is otherwise obfuscated. Picking the additional right dimension can highlight a solution that might otherwise be invisible. Many years ago, an easily stated mathematical problem was posed; it had to do with a tiled board and a set of dominos. The board had 64 equal squares arrayed in 8 rows and 8 columns, as shown in Figure 4-3. The mathematical problem concerned covering the board with dominos.

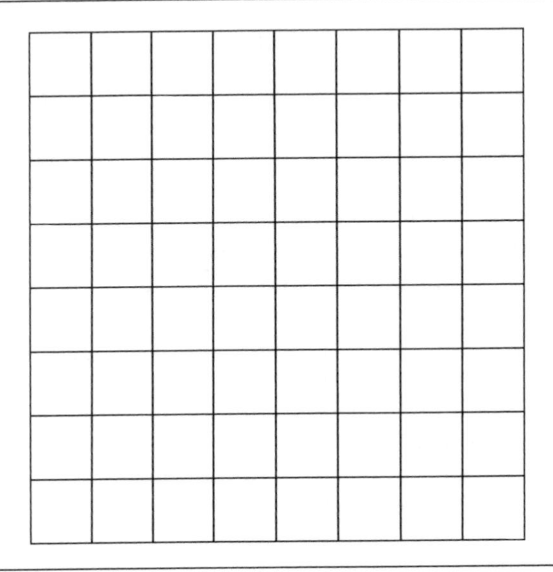

FIGURE 4-3 An 8-by-8 tiled board

Imagine that 32 identical dominos are placed on the board, as shown in Figure 4-4. Each domino is placed in a vertical orientation, and each domino covers two adjacent squares on the tiled board. The 32 dominos are placed on the board so that the entire board, all 64 squares, is covered by the dominos.

Now suppose I remove two squares from the board, specifically the ones at the upper-left and lower-left corners. Can I now cover the board with 31 dominos? The answer is yes, as shown in Figure 4-5. All I need do is place three vertical dominos over the leftmost column of six squares and fill the rightmost seven columns as before.

And now for the intriguing question. Suppose I remove two opposite squares from the tiled board, as shown in Figure 4-6. Might I now cover the 62-square-tiled board with 31 dominos?

This question bounced around academic coffee klatches for years. The problem was that there appeared to be so many possible ways of placing the 31 dominos that the combinatorics seemed overwhelming. Nobody found a solution, nor did anybody prove that a solution was possible or that a solution was not attainable. Nobody, that is, until one day, when the problem was posed to a newcomer, who, in a Eureka! moment, responded almost immediately that there was no solution. And the proof was offered by adding a dimension.

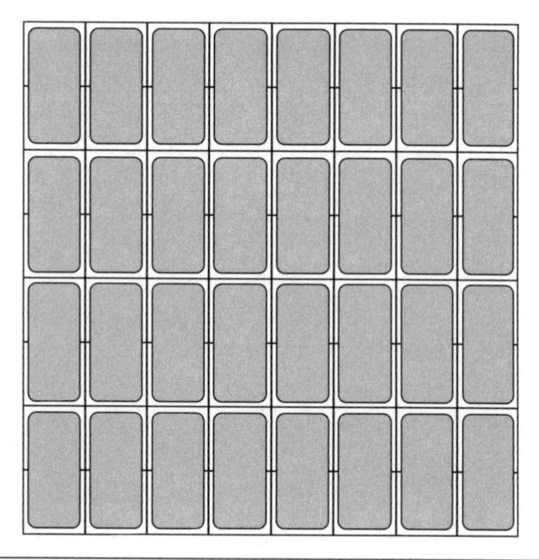

FIGURE 4-4 Thirty-two dominos cover the 8-by-8 board

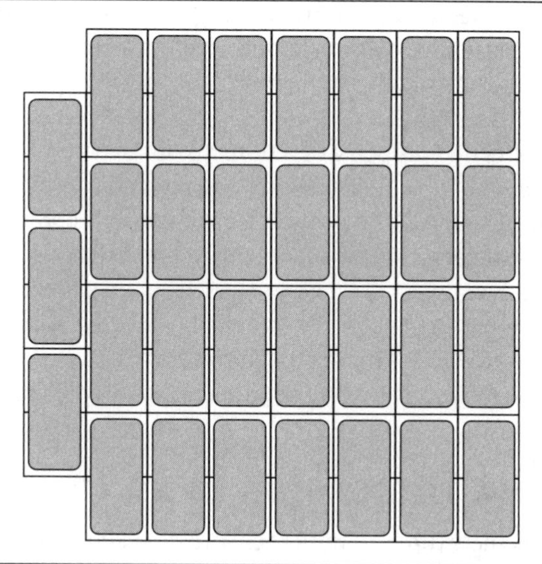

FIGURE 4-5 The board with the upper-left and lower-left squares removed

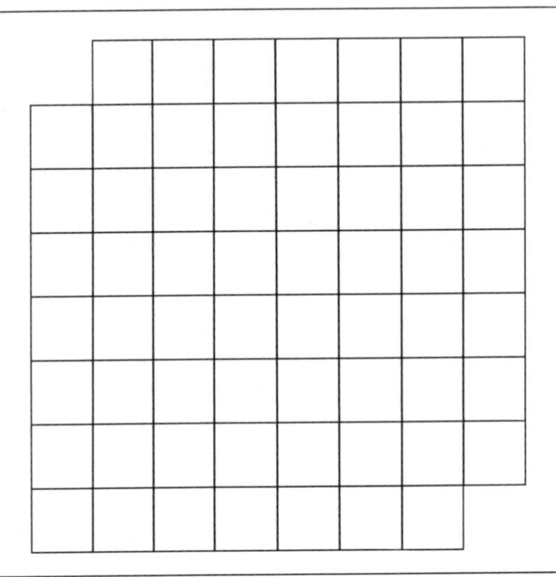

FIGURE 4-6 The 8-by-8 board with the upper-left and lower-right squares removed

Here's that proof: Think of the 64-square tiled board as a checkerboard, with 64 equal squares arranged just like the board in the example, but the squares in a checkerboard are colored or shaded: they have not only a spatial dimension, but the added dimension of color or shading, as shown in Figure 4-7. There are two colors, often red and black, or shadings, and the colors or shadings of two adjacent squares are different.

Now imagine the property of a domino. It covers two adjacent squares, and therefore each domino covers exactly one red and one black square, or one clear and one shaded square. If the color of the upper-left corner square that I removed is red, what is the color of the other square that I removed? It is also red. Thus, our 62-square board has 32 black or shaded squares and 30 red or clear squares, and there is no way that such a board can be covered with 31 dominos so that all squares are covered. This is a powerful and effective mathematical argument achieved through simply adding the dimension of color or shading.

Adding a New Dimension to an Old Space

Sometimes the new dimension of technology is added on to an older invention. Instead of the inventor making an improvement to an older invention, he creates an entirely new white space. A good example of this is found in the LED traffic

FIGURE 4-7 A checkerboard

lights discussed in Chapter 2. The improvement invention in this case was the replacement of the incandescent lamps with a group of LEDs. Because the LEDs can be turned on and off much faster than incandescent bulbs, an LED traffic light can also be used to communicate to drivers, and this is precisely what was offered in a paper and patent application ("LED Traffic Light as a Communications Device," Pang et al., Proceedings International Conference on Intelligent Transportation Systems, 1999, pp. 788–793) and is an example of adding a new purpose to an old invention:

> *The visible light from an LED... traffic light can be [intensity] modulated and encoded with information. Hence it can be used for the broadcasting of audio messages or any traffic or road information. Essentially, all LED traffic lights can be used as communications devices... in which one or more LEDs are modulated and encoded with audio messages.*

The traffic light is of course visible to a vehicle, and if that vehicle is equipped with a [direct detection] optical receiver, it would be possible to inform the vehicle and its occupants of all sorts of useful data. In a patent application filed in 1999, the inventors suggest that "Several applications of the general-purpose system are provided, including a vehicle speed limiting application, a vehicle location and guidance system application, and a portable traveler information and location system application." So the improvement invention of an LED traffic light becomes a *combination invention* if augmented by this new dimension of service.

Choosing the Right Dimensions

When we started this chapter we promised we would concentrate not only on spotting different types of dimensions, but also on selecting appropriate dimensions that contribute value. Adding dimensions without thought can actually reduce the insight to a problem. Consider the two farmers who each had a horse, and the horses were identical. They kept their horses in adjoining pens. Every so often, one of the horses would jump into the other's pen and the farmers would have to separate them. But the farmers could not differentiate which horse was which. They tried a number of ideas to tell them apart. They tied a rope collar about one of the horses, figuring that would enable differentiation. But when the horses got together the next time, the uncollared horse chewed the collar off the other. So the farmers tried clipping one horse's ear, but when the horses got together again they got into a tiff and ended up biting each other's ears and obliterating the distinguishing mark. Finally, in desperation and after much thinking, the farmers decided to measure the horses' heights. And sure enough! They found the white horse stood a full inch taller than the black horse!

The farmers were considering many dimensions, but obviously they weren't focusing on the right one.

In a more serious vein, consider the piloting of an aircraft. Long, long hours are spent by pilots in a small cabin in a generally soporific environment. But occasionally, the pilot's ability is taxed and the situation requires outstanding reflexes, magnificent mental acrobatics, and knowledge of the appropriate dimensions for salvation. One such moment occurred on January 15, 2009. It was then that Captain Sullenberger flew an Airbus A320 into a flock of geese after taking off from New York's LaGuardia airport. The engines failed and the plane started to lose speed and altitude. In testimony before the National Transportation Safety Board, Captain Sullenberger recounted that before he chose LaGuardia Airport for a return, he had to be sure he could make it there. Concisely stated, he said "I couldn't afford to be wrong." What ensued was his selection of the Hudson River as the landing area. It was reachable and, in his words, it was "long enough, wide enough and smooth enough," which were the three correct dimensions required for his successful life-saving choice.

Combining Dimensions: Considering Climate in Risk-Based Pricing

To get ready to devise combination inventions, you must first acquire a notion of what features, or dimensions, you need to bring together to create something of value that is greater than the sum of its parts. The challenge is to identify and isolate the additional one or two dimensions that are most meaningful to your effort. Quite often there are so many candidates that it is difficult and frustrating to winnow the pile of candidates down to those few valuable dimensions. This is especially true where complex effects or phenomena are functions of many different variables. There are statistical analysis packages that may help you cut through this analysis but often a consideration of some simple physics, or even what we call "common sense," can provide very helpful insight.

Some of my colleagues and I realized this when we worked on developing a method to price an automobile warranty insurance policy. Pricing of such a policy usually involves estimating the lifetime cost of the policy for a particular vehicle, and this entails knowing the car make, model, year, and the mileage, if used, when applying for a warranty.

In risk-based pricing of vehicle warranty insurance policies, the bottom line is cost to the insurer. It is a challenging business dynamic because an expected cost determination may not spread insurance costs uniformly amongst its customers resulting in low-risk customers being overcharged and high-risk customers being undercharged. The market response is that the population of high-risk customers will increase because they are undercharged and the population of low-risk customers will decrease because they are overcharged. In order to remain competitive, the vehicle warranty insurer must try to readjust the prices charged to compensate for the undercharged high-risk customers and this requires the cost model to accurately assess and weight the individual risks. This is a very complex effort because it includes many variables: the driver and his record; the variables of car make, model, mileage, and year; plus a host of environmental factors, such as mean temperature.

In the course of our study we noted an apparent yearly spike in the frequency of repairs to a car's interior climate and comfort (ICC) system. A multivariate analysis indicated the ICC's sensitivity to temperature. Looking back, this is not surprising as the ICC includes air conditioning which has to work harder and longer if the mean temperature is higher; but when you are working with many variables, it can be a bit of a jungle at first.

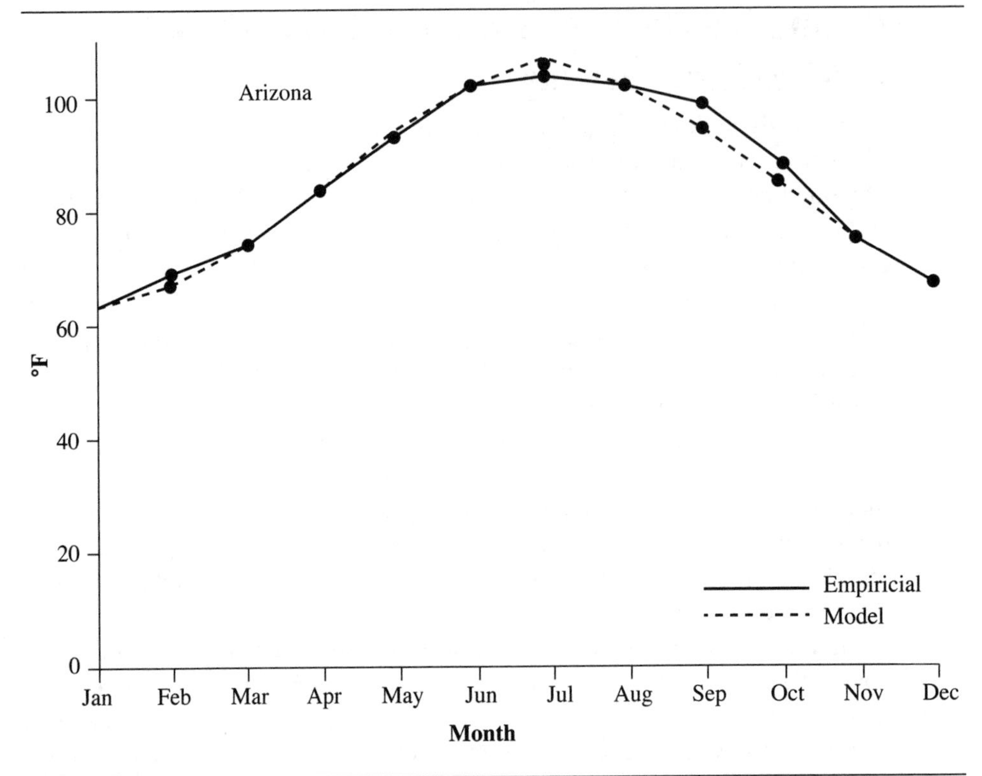

FIGURE 4-8 Some insurance claims and their correlation with mean maximum environmental temperature

Just how sensitive was the ICC to temperature? The graph in Figure 4-8, taken from our U.S. patent 6182048, answers this question with a resounding "very!" It is a plot fitting a simple one-variable temperature-based repair-prediction model to the empirical percentages of ICC claims for Arizona. Our Eureka! moment was that environmental temperature was a dimension that clearly needed incorporation into the pricing model.

Changing a Dimension: Visualizing Speech

Sometimes, instead of adding a dimension, we can gain tremendous insight and invention opportunities by simply *changing* a dimension. Sometimes the best way to understand an art is to look for its limitations: what it *can't* do, and where it is incomplete. In this way, we can seek out a new dimension of understanding to help spark inventive ideas.

Consider human speech again. Although we use speech as our main method of communication, we know that speech has limitations. For one thing, words used in speech can be subject to interpretation. For example, without context, when we speak the words *to*, *too*, and *two*, they sound the same, but their meanings are very different, of course. I went *to* school. I went, *too*. *Two* of us went to school.

But speech can be considered in other dimensions as well, and by thinking in these dimensions, we can come up with potential for innovation and invention.

We can visualize speech, for example. Figure 4-9 shows an example of a waveform of a speech signal. It shows the time-varying voltage produced by a microphone as a speaker spoke a sentence. We'll get back to this in a minute. In this waveform, notice that the voltage appears to be a complex waveform but alone, without interpretation or more information, we cannot tell what is being said, even though the waveform shows us that something is being spoken. The immediate question for the inventor is how to turn the waveform into something more intuitive to the human eye-brain computer.

When we analyze human speech, we see that it is a complex and coordinated process that involves the vocal cords and manipulations of the mouth, tongue, and lips. We humans use three basic types of sounds in speech.

The first is produced when our vocal cords vibrate and excite resonances within our vocal tract. These sounds form *vowels*. A vowel sound has a very distinct structure in that it consists of modulated tones, or *formants*, over a short period of time.

Another speech sound is called a *fricative*. This is an unvoiced sound, when the vocal cords are not vibrating, and the air is just rushing past them—they sound like random noise, such as the *f* in *f*our and the *s* in *s*ix. A fricative has a broad energy spectrum—that is, it exhibits energy across a wide range of frequencies.

The third sound type is the *plosive*. This is produced by a temporary closure followed by a rapid opening of the vocal tract as air goes past the nonvibrating vocal cords. This sound is used in producing a *d*, a *t*, and a *p*. A plosive is a very

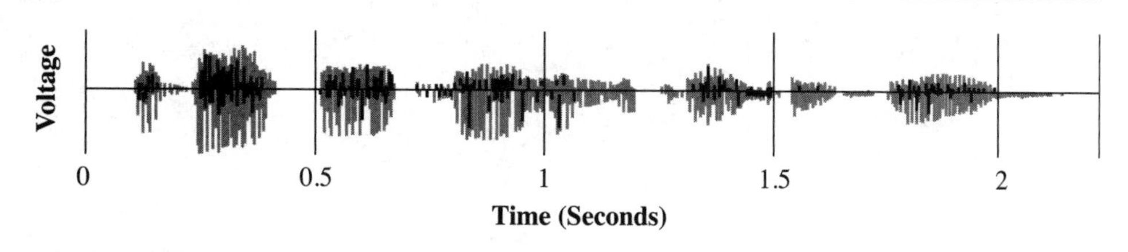

FIGURE 4-9 Voltage waveform of a speech signal (Courtesy Prof. J. Picone)

short duration signal. It is unlike the fricative in that it persists for only a short time, but like the fricative, it exhibits a broad energy spectrum.

If we had some way, a dimension other than voltage, to present the activity within the frequency spectrum, we might be able to create a better way of visualizing speech. This swap of dimensions, what is called a "short-term frequency power spectrum for the voltage dimension," is precisely the basis for a strategic invention for visualizing speech that was developed during World War II by R. K. Potter and colleagues at Western Electric. A device to display the short-term frequency power spectrum was developed and called the "sound spectrograph" and the display was called a "spectrogram."

Let's look at a few simple spectrogram examples to show you how it works. The first shows a series of piano notes shown in Figure 4-10. (The spectrograms shown in Figures 4-10, 4-11, 4-12, and 4-13 are from Summary Technical Report of Division 13, National Defense Research Committee [NDRC], Vol. 3, *Speech and Facsimile Scrambling and Decoding*, Washington, D.C., 1946, [declassified, 2 August 1960]). Remember that, in the spectrogram, time is on the horizontal axis and frequency is on the vertical axis; the higher the frequency, the higher its value on the vertical axis. Notice on the spectrogram that the piano music consists of a sequence of many tones, all beginning at the same time.

Now take a look at the spectrogram for random noise, or white noise, shown in Figure 4-11. White noise has no persistent tones, no clearly defined starts or stops. Instead, the white noise spectrogram shows energy randomly distributed throughout the time-frequency space.

Finally, let's look at a spectrogram produced on the sound produced by crumpling a piece of paper, in Figure 4-12. This spectrogram shows a variety of

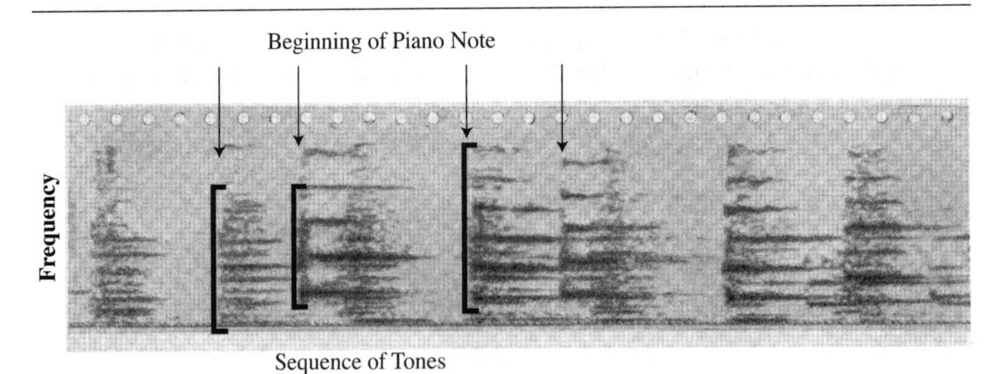

FIGURE 4-10 A spectrogram of piano notes

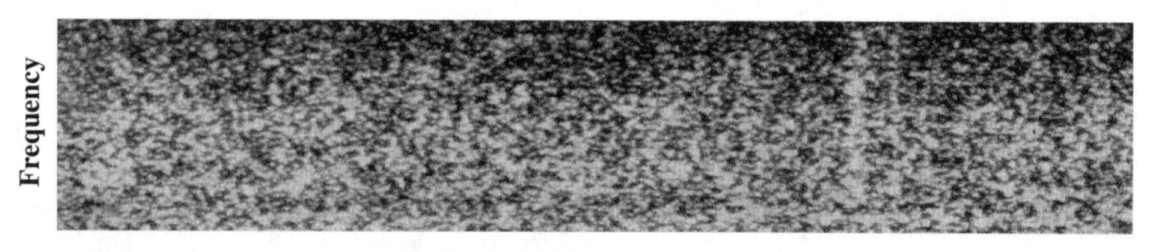

FIGURE 4-11 A spectrogram of white noise

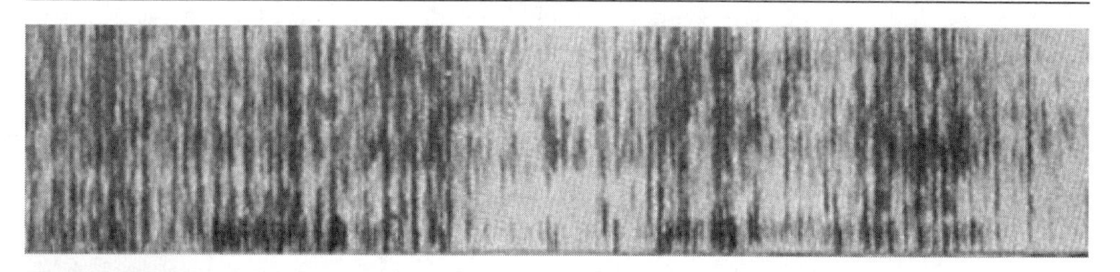

FIGURE 4-12 A spectrogram of the noise produced by crumpling paper

effects. First, imagine what crumpling a paper sounds like. It's a series of crackles or pops, irregularly spaced in time. These crackles or pops are sometime referred to as *impulses* and their spectrum shows a very broad band of frequency, like random noise.

Now, let's get back to the dimensionality of speech. Remember that voice has three components: voiced sounds with their formants, unvoiced hissing termed fricatives, and plosives, broadband signals of very short duration. We can overlap the corresponding spectrogram (courtesy of Prof. J. Picone) shown in Figure 4-13, the alternate dimension, above the time trace of the voice signal that we first introduced.

In Figure 4-13, note that the fricative *th* spreads its energy over a wide range of frequencies. The formants of the first *o* in doctor follow the plosive *d*, and then a break, a silence, occurs between the end of the first syllable of doctor and the second syllable, which also starts with the plosive *t*.

Speech, when viewed in a spectrogram, has a lot of linkage or structure, and a lot of redundancy. This is the structure that made analysis of WWII analog speech scramblers possible. One speech scrambling technique cut a speech

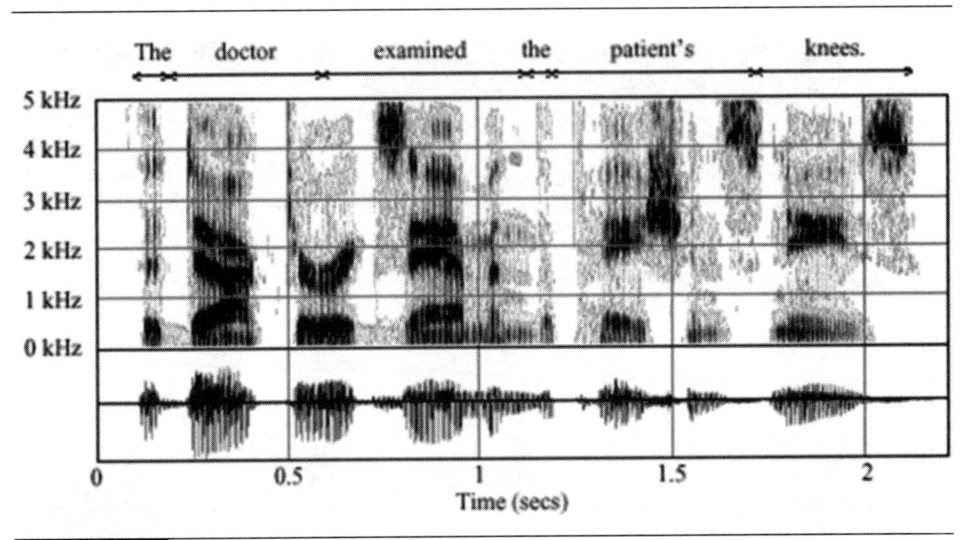

FIGURE 4-13 The spectrogram corresponding to the voltage waveform of Figure 4-9

signal into time-bandwidth blocks and randomly shuffled them, as shown in the top spectrogram of the spectrogram pair shown in Figure 4-14. By using a spectrograph and matching formants and other elements of the scrambled speech, it was possible for analyzers to piece together the unscrambled speech, as shown in the bottom spectrogram of the figure.

The NDRC report states that the fact that "both time and frequency variations are simultaneously displayed which makes spectrograms so valuable for decoding work." Think a bit: Could any of this value be harvested for intellectual property? Might this changing of a dimension nurture a Eureka! moment?

The answer is decidedly Yes! Sighted deaf persons might be able to use the spectrograph interpretation to understand speech that they cannot hear. Consider the following excerpt (from Chapter 7, "When Will HAL Understand What We Are Saying?" Computer Speech Recognition and Understanding HAL's Legacy, Raymond Kurzweil, of *HAL's Legacy*, MIT Press, 1996):

> *The device [the telephone] ultimately broke down the communication barrier of distance for the human race. Ironically, Bell's great invention also deepened the isolation of the deaf. The two methods of communication available to the deaf—sign language and lipreading—are not possible over the telephone.*

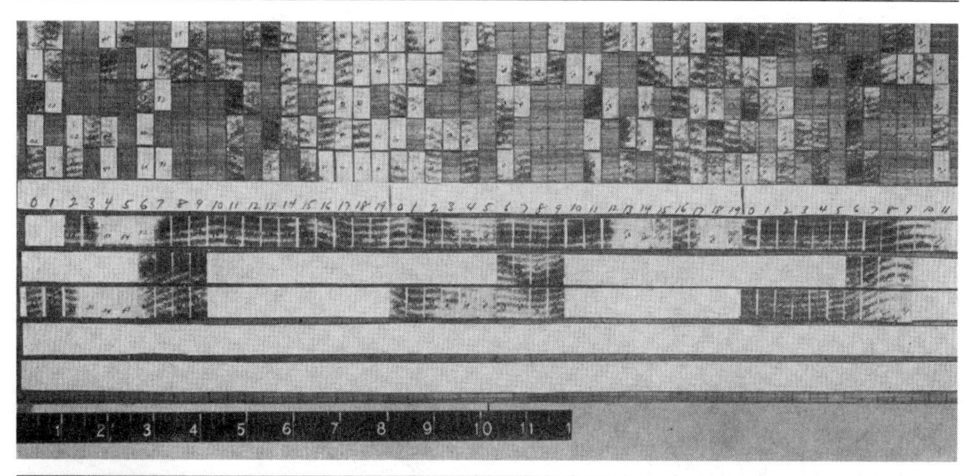

FIGURE 4-14 Pseudorandom shuffling of frequency-time blocks of speech and its reconstruction

Enter one embodiment of U.S. Patent 5532936 "Transform Method and Spectrograph for Displaying Characteristics of Speech," which addresses the use of a spectrograph for just this purpose. It can be understood from figure 7 of the patent, shown in Figure 4-15, and the following paragraph from the description:

> *One embodiment of the present invention represents a telephone signal in graphical format. The telephone signal which is carried as an alternating current signal between telephone lines Ring 62 and Tip 64 are processed by realtime spectrograph 66 and displayed on spectrographic display 68, enabling a deaf user to see the speech of all parties on the line, and enabling effective use by said deaf user of ordinary telephone set 70. An alternate embodiment of the present invention incorporates a spectrograph and display within a telephone set.*

This example addressed the issue of *sensor compensation*. Lacking an audio channel to the brain, how might we use other sensors such as our eyes to gain the sense of a spoken message? Specifically, how do we process the incoming data so that it can be interpreted when displayed on a flat, two-dimensional screen? Any processing and display that will lead to a faster understanding of the data is likely to have many uses. The challenge to the inventor was to find the useful dimensions

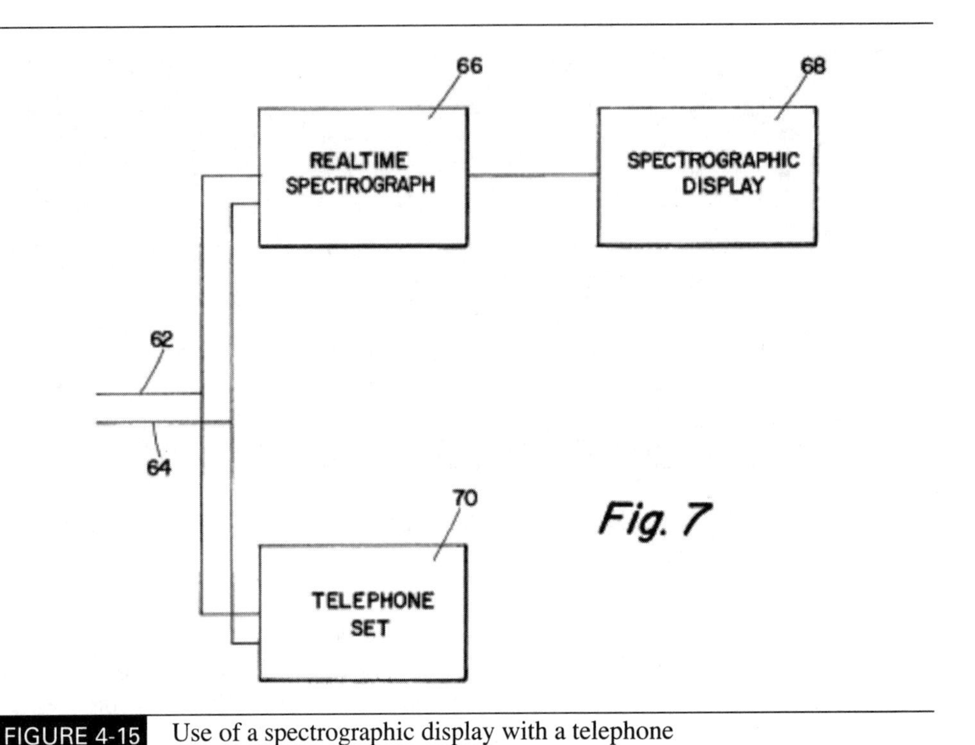

FIGURE 4-15 Use of a spectrographic display with a telephone

for preprocessing the data and then inventing or adapting an optimal visual format for displaying the preprocessed data to our human eye-brain computer.

Recap

In this short chapter, you learned that the incorporation of extra dimensions or features to a product or process can result in something of greater value than the sum of its parts. New dimensions must be chosen carefully so that they will provide increased value. In addition, you saw how changing a dimension can result in significant value for the inventor. These topics should help expand your inventive vision. They are expected to aid in positioning you better to practice the art of combination inventions, which is the focus in the next chapter.

Discussions and Reflections

- Select some successful inventions and analyze them to determine what features or dimensions they bring together. Try to estimate the value of each feature independently and see if the sum of the features or dimensions has resulted in an invention whose value is greater than the sum of its parts.

- General Ludwig Beck was one of the plotters against Adolf Hitler in Operation Valkyrie during World War II. Much earlier, Beck directed a group of officers in preparing a famous military work on troop command (*Truppenführung*). In this work is a contribution from a General Kurt von Hammerstein-Equord (from http://en.wikipedia.org/wiki/Hanlon%27s_razor):

  I divide my officers into four classes: the clever, the lazy, the industrious, and the stupid. Each officer possesses at least two of these qualities. Those who are clever and industrious are fitted for the highest staff appointments. Use can be made of those who are stupid and lazy. The man who is clever and lazy however is for the very highest command; he has the temperament and nerves to deal with all situations. But whoever is stupid and industrious is a menace and must be removed immediately!

  Reflect on the people you've known and worked with. Do they have multidimensional attributes that promote inefficiency or perhaps annoyance? Instead of people, how about products? Situations? Policies in general?

  Consider also a lack of dimensional attributes espoused by the following Japanese proverb: "Vision without action is a daydream. Action without vision is a nightmare."

Chapter 5

Combination Inventions

To invent, you need a good imagination and a pile of junk.

—Thomas Edison

A combination invention results by melding two or more exciting methods or technologies into a novel system. Patentable new inventions can be conceived by bringing together disparate elements to form a new and useful combination that is nonobvious. In previous chapters we discussed ways of looking at the many dimensions of a device or system. A useful way to increase the dimensions is to bring together combinations of things that might at first seem unrelated.

It is possible for a combination invention to result in something of immense value by incorporation of a new feature or nuance of one of the components being combined. The combination invention might look and seem to behave exactly like one of the original parts, but buried within it is a powerful new feature. These hidden features may include such things as greater resistance to problems plaguing the original version of the device, better packaging, more efficient algorithms, longer operational life, and clever and subtle redundancies.

A Very Large Scale Combination Invention

During the Cold War, it seems that the Soviets understood the concept of combination inventions quite well, and it spilled over into their political military policy on intercontinental ballistic missiles (ICBMs). The United States put its land-based missiles under the control of the Air Force, an existing arm of the military. But the Soviets did not follow suit; instead, they created a completely new service, the Strategic Rocket Forces. The reason, it is thought, is that the Soviets recognized that the ICBM represented a combinatorial, or in Marxist terms, a qualitative innovation in weaponry, for it was a new system founded on the synergy of three different components: the missile, the nuclear warhead, and the guidance and control system.

Combination Invention and Emerging Technology

As different technologies emerge and mature, an innovator may correctly sense the value that would be unlocked by combining pieces of the different technologies. It is sometimes surprising how compelling such a combination appears in hindsight. Once realized, the combination may acquire an almost viral nature and inventions begin to pour out supporting the combination. We can see clear and compelling evidence through simple patent statistics just how significant, and how infectious, a particular combination invention can be. A great example is provided via the two seminal technologies: the Internet and the digital camera.

The Web and the Camera

Table 5-1 looks at the number of patents awarded these two technologies that were destined for combination. The data spans 13 years (1990–2002) and enumerates the number of issued U.S. patents that include the terms "Internet," "digital camera," and "Internet" and "digital camera" in their claims, title, or abstracts. You can see that it didn't take long for the combination inventions to break loose.

The first combination invention patent, U.S. Patent 5845265, "Consignment Nodes," was filed in 1995. Take a look at the abstract:

A method and apparatus for creating a computerized market for used and collectible goods by use of a plurality of low cost posting terminals and a market maker computer in a legal framework that establishes a bailee relationship and consignment contract with a purchaser of a good at the market maker computer that allows the purchaser to change the price of the good once the purchaser has purchased the good thereby to allow the

Filing Year	Internet	Digital camera	Internet and digital camera
2002	17861	2620	147
2001	14978	2039	120
2000	11081	1463	67
1999	7031	950	37
1998	4225	621	21
1997	2297	337	10
1996	874	166	2
1995	199	84	1
1994	47	47	–
1993	23	35	–
1992	14	29	–
1991	8	23	–
1990	8	18	–

TABLE 5-1 Patents Including the Terms "Internet" and "Digital Camera"

purchaser to speculate on the price of collectibles in an electronic market for used goods while assuring the safe and trusted physical possession of a good with a vetted bailee.

And take a look also at three of the claims, numbers 1, 4, and 7:

1. A system for presenting a data record of a good for sale to a market for goods, said market for goods having an interface to a wide area communication network for presenting and offering goods for sale to a purchaser, a payment clearing means for processing a purchase request from said purchaser, a database means for storing and tracking said data record of said good for sale, a communications means for communicating with said system to accept said data record of said good and a payment means for transferring funds to a user of said system, said system comprising:

- *a digital image means for creating a digital image of a good for sale;*
- *a user interface for receiving textual information from a user;*
- *a bar code scanner;*
- *a bar code printer;*
- *a storage device;*
- *a communications means for communicating with the market; and*
- *a computer locally connected to said digital image means, said user interface, said bar code scanner, said bar code printer, said storage device and said communications means, said computer adapted to receive said digital image of said good for sale from said digital image means, generate a data record of said good for sale, incorporate said digital image of said good for sale into said data record, receive a textual description of said good for sale from said user interface, store said data record on said storage device, transfer said data record to the market for goods via said communications means and receive a tracking number for said good for sale from the market for goods via said communications means, store said tracking number from the market for goods in said data record on said storage device and printing a bar code from said tracking number on said bar code printer.*

4. The apparatus of claim 1 wherein said image input means is a digital camera.

7. The apparatus of claim 1 wherein said communications means is via an internet.

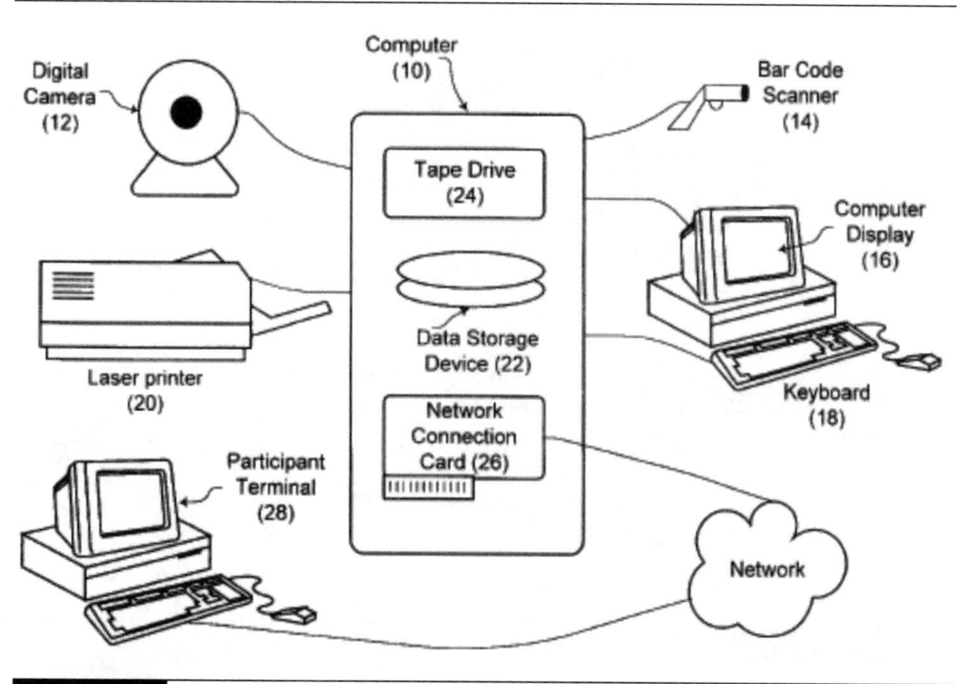

FIGURE 5-1 Figure in U.S. Patent 5845265

Figure 5-1 shows a figure included in the patent. You can see the utility generated by combining the two different technologies of the Internet and the digital camera. Just as television incorporated visual imagery with voice, the combination of the digital camera and the Internet combines two essential modes of communication: text and image. The value of such a combination truly exceeds the sum of its parts.

Bar Codes and Cooking

Let's look at one more example and this time with two highly dissimilar technologies, bar codes and cooking. Is this an unlikely place to find white space?

It turns out that a lot of innovation links these two technologies. A great example is illustrated in U.S. Patent 6862494, "Automated cooking system for food accompanied by machine readable indicia." The abstract and the patent's figure 1 (shown in Figure 5-2) and parts table (shown in Table 5-2) are intriguing.

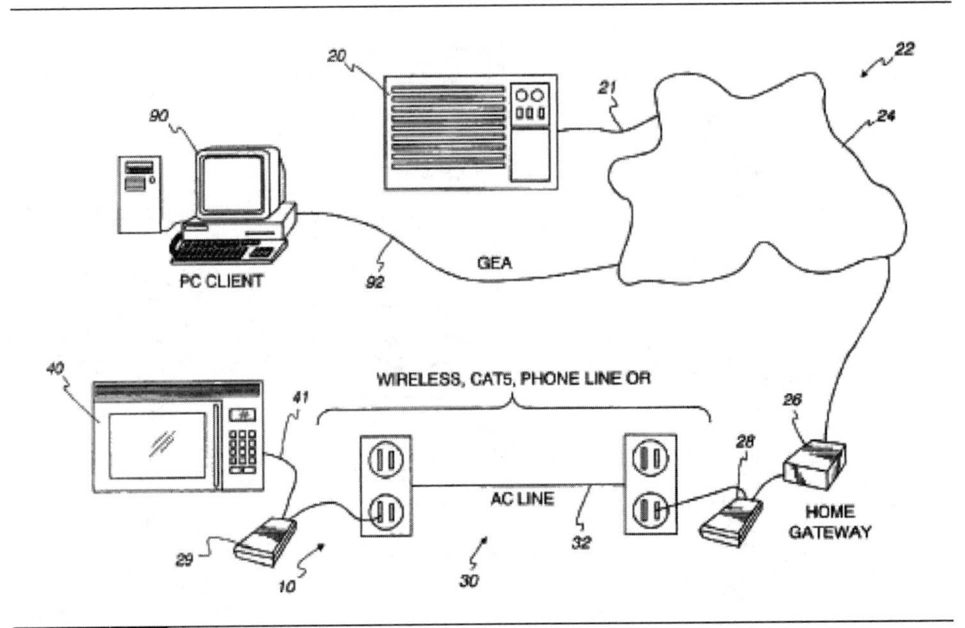

FIGURE 5-2 U.S. Patent 6862494's illustration

Component number	Description
10	cooking system
20	Internet server computer
21	cable
22	a network such as the Internet including 24
24	a wide area network
26	home gateway
28	modem
29	modem
30	a local area network
32	AC power line
40	oven
41	cable
90	personal computer
92	cable

TABLE 5-2 Patent 6862494 Parts

An automated cooking system... cooks food... accompanied by machine-readable indicia, such as a bar code... read by a reader.... Cooking data, including function coefficients, are accessed from an Internet server... based on the information in the bar code. An oven... cooks the food in phases in response to the cooking data and one or more monitored food parameters of humidity, temperature and weight. The cooking is controlled by functions in which one or more food parameters are multiplied by coefficients that vary according to food type.

This patent offers just a small hint of the breadth of combination inventions. Thanks to the World Wide Web and wireless technologies, linkages are developing between more and more devices and functions and, as you shall see later on in this chapter, barriers to combination inventions continue to diminish.

An Unusual Combination Invention

As a nation, one of our favorite animals is the domestic cat. In the 1990s, 19 U.S. patents were issued involving cat food or a cat feeder. I recently discovered a rather unusual combination invention that I can't resist sharing. U.S. Patent 4150505 has the grim title of "Bird trap and cat feeder." The inventor was clearly convinced that some birds were a distinct problem to society, and that hungry cats could be the solution to the problem. Here is an excerpt from the invention's summary:

The subject invention solves the problem of reducing the population of sparrows.

Heretofore, the ordinary sparrow has greatly proliferated, thereby taking the place of more popular birds, such as the canary, blue bird, wren, swallow, and other birds that are appealing to the eye and enjoyable to listen to. Also, because of the increased population of the sparrow, the bird has become a nuisance due to bird droppings, the building of nests, and the taking of food supplies which would ordinarily be enjoyed by other birds.

The invention provides means for continuously trapping sparrows and supplying a cat and neighborhood cats with a supply of sparrows. The cat feeder by its design is self-cleaning since the cat quickly learns to remove the sparrow from the cage.

(continued)

The trap provides a weighted balanced perch, which simulates a tree branch and lulls the bird into believing he can fly outwardly from an opening in a bird housing which is covered with a glass window.

The bird trap uses an elongated down spout which is attached to the bottom of the bird trap and the top of the cat feeder so that the bird trap may be placed at the top of a building structure while the cat feeder is placed near the ground level so that the cat may remove the bird therefrom.

The bird trap and cat feeder includes an enclosed bird housing having an upper portion, a tapered lower portion, a back, a front, a top, an open bottom, and sides. The housing includes entrance holes in the sides of the upper portion of the housing for receiving the bird therein. A balanced perch is pivotally mounted inside the housing and adjacent the entrance holes. A glass window is mounted in an opening in the tapered lower portion of the front of the housing. The cat feeder includes a one inch wire mesh cage which is communicably connected to the open bottom of the housing for receiving the bird therein."

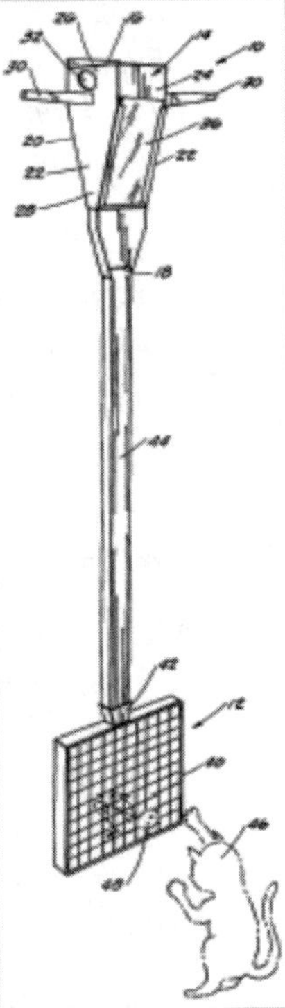

In the illustration, the patent makes the concept a bit clearer (46 is the cat; 48 the sparrow). The inventor added a dimension—he added the cat feeder to the bird trap. The invention rose to a patentable invention exhibiting utility, novelty, and nonobviousness. Although it is certainly an original combination invention, in my opinion it is also a somewhat disturbing one; I am thankful it seems not to have caught on.

Subliminal Channel Concept: Locks and Alarms

The subliminal channel concept is present in a number of instantiations. The concept is quite powerful and in one case demonstrates the value of a single bit of information. The subliminal channel can often be used to game the system. The following example illustrates such a gaming wherein the "system" is a security container that may be opened in two very similar ways but with very dissimilar results. You will see that the combination invention results from combining a conventional safe and alarm system with the subliminal channel.

The combination lock was invented more than 800 years ago by an Arab scholar named al-Jazari. Today combination locks are a standard of physical protection for many secure containers. What are the dimensions of a physical combination lock?

- The combination numbers, which are usually three but up to six for some locks
- The order of the numbers—the sequence in which they are dialed
- A third dimension, one not usually considered: the initial direction in which the combination dial is turned—for example, clockwise or counterclockwise

It is a largely unknown property of many high-end safes, but the safe can be opened on correct entry of the combination sequence *regardless* of the initial direction in which the combination dial is turned. This third dimension is an additional bit of information, and it can be put to great use.

Most businesses instruct personnel to comply with an armed robber's demands to avoid injury. For our example we look at the case wherein an establishment's safe with a combination lock is targeted. The implied threat is lethal force against a noncooperating employee.

As depicted in Figure 5-3, the clerk is told to open the safe and, of course, not activate an alarm. The combination to the safe is 10-20-30. If the clerk enters the combination with the initial direction being a counterclockwise rotation of the dial, the safe will open and it will not trigger the silent alarm. If, however, the clerk starts in the clockwise direction, the safe will open and apparently comply with the robber's demand, but a silent alarm will be triggered.

Now, if the robber were well read and knew of this phenomenon, he might order the clerk to start in the counterclockwise direction. But, perhaps the counterclockwise direction had been chosen beforehand and set as the alarm condition? There might be no way for the robber to know. The carrying of the extra bit of information in the entry of the combination sequence is an instance of what is known as a subliminal channel, so named by Gus Simmons, then of Sandia National Labs in Albuquerque.

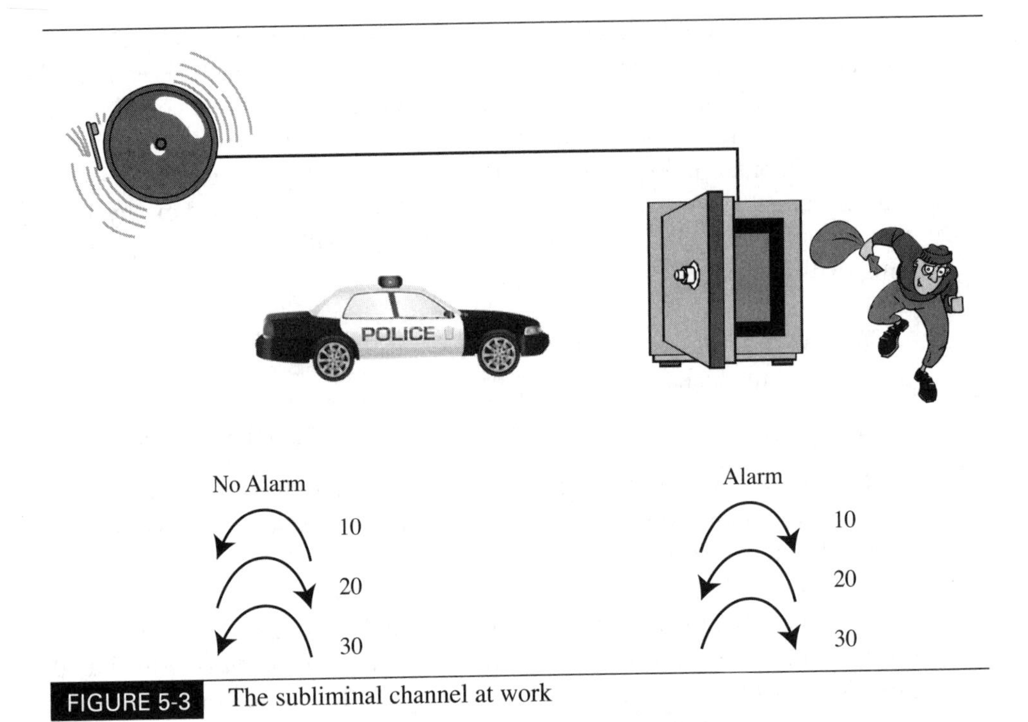

FIGURE 5-3 The subliminal channel at work

The concept and its use extend naturally to more modern locks, such as the ten-button keypad shown in Figure 5-4. Similar to the combination lock, the dimensions include the numbers and their correct order, but there is no clockwise/counterclockwise gambit. There is, however, the possibility of entering a different combination and reaping the benefit of a subliminal channel. For someone to access an area, they might have to enter 1, 2, 3, 4. If they enter instead 1, 2, 3, 5, they will also be accommodated access but that combination will trigger the silent alarm.

A Wartime Countermeasure

In *As You Like It*, Shakespeare gave the enduring byword that "one man in his life plays many parts." When it comes to inventors, Shakespeare had it half right, for *women* also play many parts.

Hedy Markey was such a woman. A film star in Europe in the 1930s, she was married to an industry magnate and living a very comfortable life as she practiced her art of avant-garde filmmaking. Things might not have developed beyond that

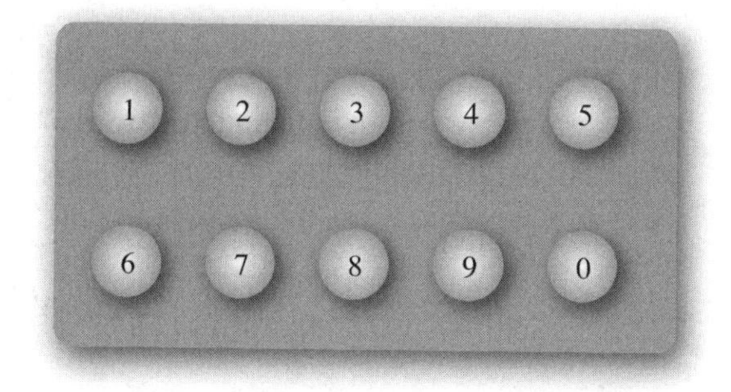

FIGURE 5-4 Ten-button keypad lock

Adding Third Dimension

The mission for the serious combination innovator is to find a way to game an old system by touching it on its insides. And this can be done only by adding a new dimension to the old game—making the old game new by playing a new part in the old game.

Back in 1884 a distinguished educator and theologian, Edwin Abbott, published an intriguing tale titled "Flatland: a Romance of Many Dimensions." Set in the style of a social allegory, the tale has increased in popularity over the years. It is set in two-dimensional Flatland, where the inhabitants comprise circles, squares, and other polygons. There are strict social rules for behavior motivated by concerns for safety as well as courtesy. After all, in two dimensions, inhabitants cannot easily tell the shape of another being. Quite often, they need to move about its periphery to count the number of its sides, and if, for example, a triangle were running through Flatland and one of its sharp angular boundaries were to impact, say a circle, the result could be quite painful, indeed even fatal, for the circle.

The tale becomes relevant to us at its point of transcendence. A Flatland square is visited by a super being, a sphere. How does the Flatland square classify the sphere? As the square can interact only with the portion of the sphere that is in Flatland, the square concludes that the visitor is a circle. But this new circle is very interesting. As the sphere rises and lowers relative to the plane of Flatland, the circle that the square encounters alternately shrinks and grows in size. Such dynamism in size is perhaps interesting to the square,

(continued)

but not spooky. It becomes spooky when the sphere disappears by completely rising above the plane of Flatland and then reentering Flatland on the *inside* of the square, touching it within its two-dimensionally closed innards.

> *The sphere said to the square: "Now I shall come back to you. And, as a crowning proof, what do you say to my giving you a touch, just the least touch, in your stomach? It will not seriously injure you, and the slight pain you may suffer cannot be compared with the mental benefit you will receive."*
>
> *The square related what then happened: "Before I could offer a word of remonstrance, I felt a shooting pain in my inside, and a demoniacal laugh seemed to issue from within me. A moment afterwards the sharp agony had ceased, leaving nothing but a dull ache behind, and the Stranger began to reappear, saying, as he gradually increased in size, 'There, I have not hurt you much, have I? If you are not convinced now, I don't know what will convince you.'"*

point except for three things. She had a first-rate mind, a great social conscience, and she could not abide what she perceived as a terrible threat posed to society by Adolf Hitler.

Guided by principles, she left her magical life in Austria, came to the United States, and proceeded to see how an esoteric Allied technical vulnerability concerning radio-controlled torpedoes could be countered. She was concerned that the radio links from shipboard to the torpedo could be jammed by Axis electronic warfare capabilities.

With a colleague, George Antheil, she worked out a scheme whereby the radio control link could be moved rapidly from one frequency to another to escape interference by one or more jammers. Their work was presented to the U.S. Patent Office, which initially issued a secrecy order on the invention that later issued as U.S. Patent 2292387 for a "Secret Communication System" on August 11, 1942. This patent could successfully game a deadly game, and it did so by adding some important dimensions.

Frequency Hopping

To help you understand their patent, we'll look at a typical scenario for a jamming problem, as shown in Figure 5-5. If the jamming signal is received at the same

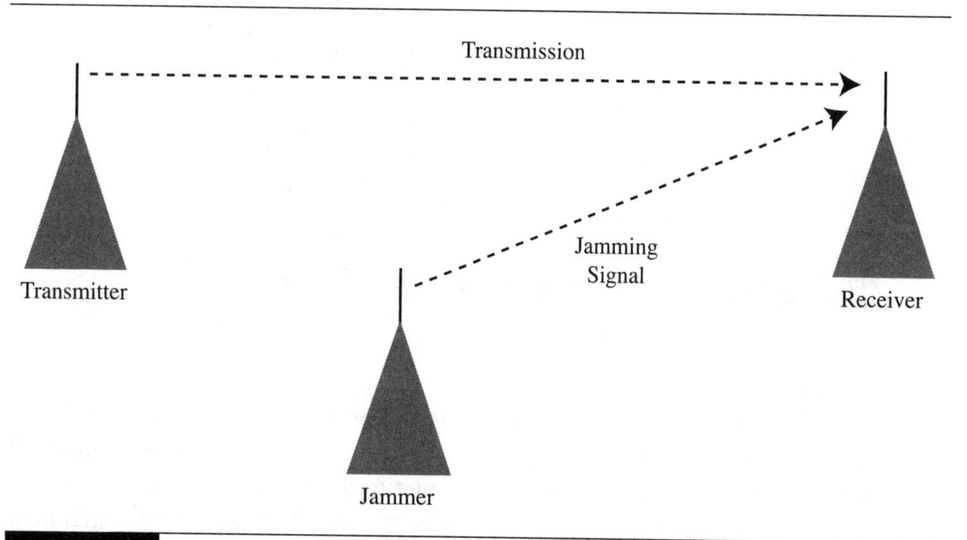

FIGURE 5-5 A jamming geometry

time as the transmitted signal and in the same frequency space as the transmitted signal, the jamming signal might interfere with the transmitted signal and make its accurate reception and proper demodulation impossible—that is, the transmitted signal is jammed.

Generally, there is a frequency band within which the transmitted signal will be transmitted; such a band is a contiguous span of radio frequency spectrum that has been allocated or zoned for a particular class of transmissions such as the AM band, which runs from 530 kHz to 1700 kHz.

Let's assume that the transmitted signal is transmitted in the frequency band, as shown in Figure 5-6.

Now let's assume that the jamming signal is laid over the transmitted signal, as shown in Figure 5-7.

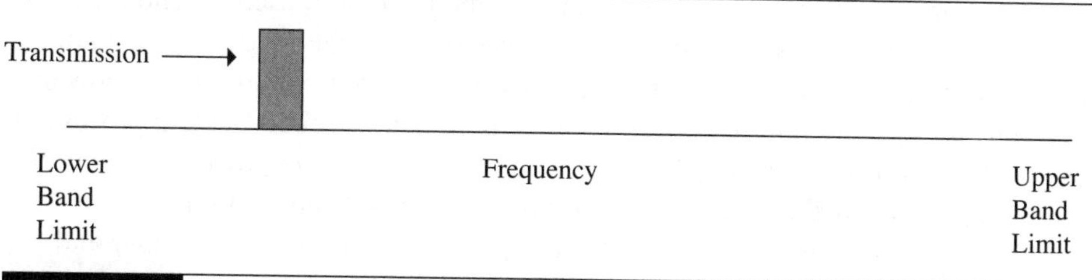

FIGURE 5-6 The frequency band for transmitted signals

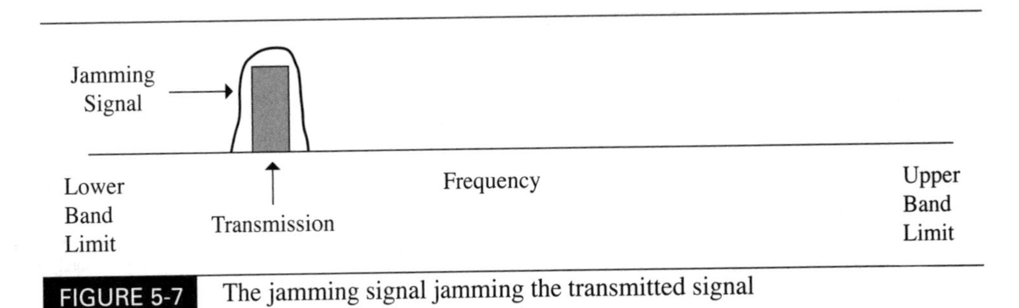

FIGURE 5-7 The jamming signal jamming the transmitted signal

To perform a successful jamming operation, the opposition locates the frequency of the transmitted signal and shifts the frequency of its jamming signal so that its jamming signal overlays the victim transmitted signal.

What Markey and Antheil taught was to move the transmitted signal in an apparently random motion within the frequency band. This way, the opposition would have to keep hunting for the frequency of the victim's transmitted signal and keep moving its jamming signal to follow and overlay the victim's signal before the victim's signal would be moved again. The victim's signal would appear to the opposition as a signal hopping around in frequency, so it is not surprising that this electronic countermeasure became known as frequency hopping. See Figure 5-8.

Frequency hopping is a great idea, and appeared to be a good countermeasure—but there is a missing element. Do you see it? The problem is that the intended receiver has to know in advance where the transmitted signal will be in the frequency band. If the transmitter is moving the transmitted signal around by using a publicly known process, then the jammer will know where the signal will be. If the transmitter is moving the transmitted signal around by using a *random* process, then the receiver will not know where to listen. So what the transmitter needs to do is move the signal around using a *pseudo-random* process that is known to both the transmitter and the receiver but not to the jammer. A pseudo-random process can be produced using well-known cryptographic techniques, and it cannot be distinguished from a true random process by someone who does not know the underlying deterministic process that is running the pseudo-random process. It is the *combination* of frequency hopping with a pseudo-random process for controlling the hopping that makes this combination invention viable and extremely valuable.

This patent is a very important one, and much of modern military and commercial communications stands on its shoulders. By the way, Hedy Markey is more widely known by her stage name, Hedy Lamarr.

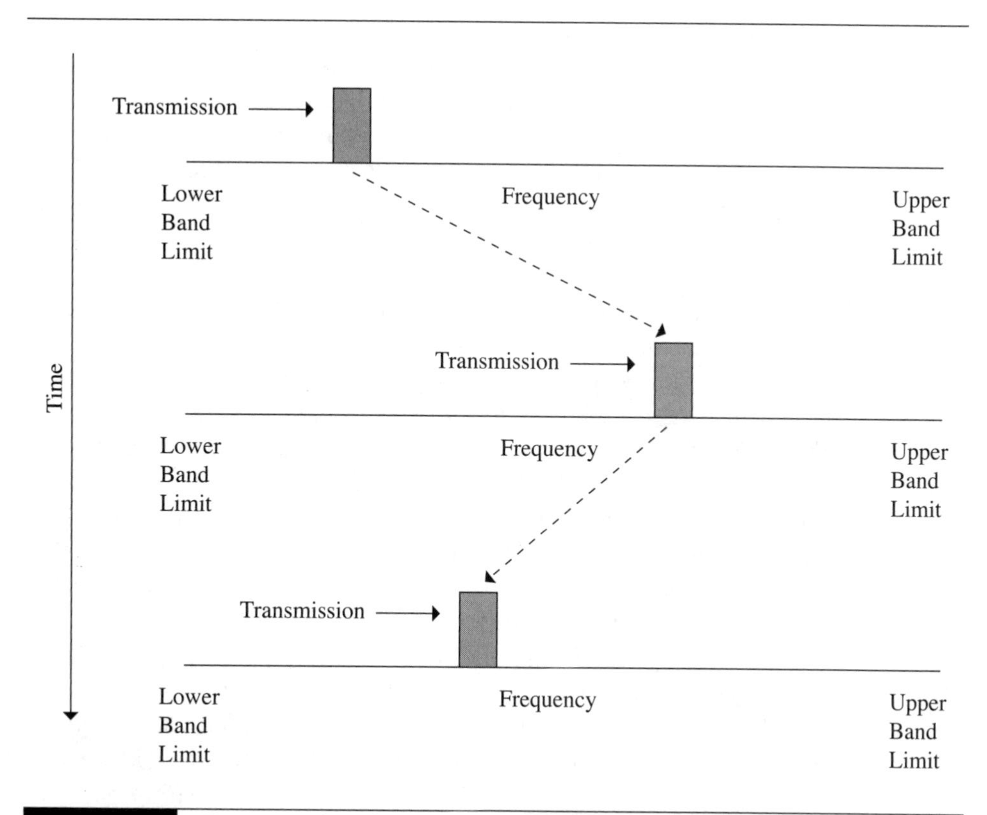

FIGURE 5-8 Frequency hopping to avoid being jammed by a follower jammer

A No-Holds-Barred Approach to Combination Invention

We have so far examined a number of examples of technologies coming together to form a combination innovation. But how might we do it? In this book's Introduction, I said that occasionally Eureka! moments can be produced by taking a *no-holds-barred approach* to a quest for solution. By this I mean that you assume that whatever technology you might need to build or enable your conceived invention can be procured or straightforwardly developed and, later on, you go back and fill in those assumptions. The invention I'm about to describe is a result of such a Eureka! moment.

The background to the specific problem facing a group of us was the question of how to monitor a large number of machines. These machines had moving parts that caused vibrations that were transmitted to the machines' housings. When a machine's vibrations became excessive, or when the vibration entered a particular mode, it was a sign that the machine's health was degrading and it needed attention.

A number of methods exist for monitoring a large group of such machines, but the operations' manager must consider cost, scale, and maintenance needs of the monitoring system itself. The background to U.S. Patent 7,667,827, "Systems and methods for remote monitoring of vibrations in machines," discusses these issues:

In order to determine the magnitude and nature of the vibration, transducers may be attached to, or placed in contact with, the machine housing so as to monitor the vibration. Subsequent analysis of the vibrational information determined by the transducers may provide valuable diagnostic data relating to the health of the monitored machinery. However, to monitor all machines, a transducer must be attached at every location in which monitoring of the vibration is desired. The health of the individual transducer also must be further monitored to ensure that the transducer is properly operating. In a facility employing thousands of machines, the purchase, installation, maintenance, and repair of thousands of transducers can become excessively expensive.

Alternatively, portable data collectors (PDC's) may be used to monitor the health of the monitored machinery. The PDC's are mobile devices that include a transducer and recording means to monitor the health of the machines. The PDC is transported to a machine and the transducer is placed in contact with the housing of the machine. The PDC can then record the vibrations of the housing. These recordings are subsequently analyzed in an attempt to assess the health of the machine. Operators must personally visit each machine and operate the PDC to record the vibrational information. The PDC operator may, therefore, spend an exorbitant amount of time monitoring the machines.

These current methods of monitoring the health of machines generally are expensive and time consuming. Often, a machine is in good health and does not require an extensive diagnostic check up. On the contrary, a failing machine may require more frequent data analyses. Thus, significant resources may be wasted. Therefore, there is a need in the art for a system and method for efficient remote monitoring of the vibrational health of a machine.

Our group gathered to see if we might find another approach to monitoring a large group of machines. We knew enough physics to realize that if a beam of electromagnetic energy, such as a laser, were shone on a machine's housing, the housing vibration would modulate the frequency of the light scattered off the machine's housing, and that the vibration could be analyzed from the scattered light. But before this technique could possibly be practical, we had to make some assumptions.

First, the source of the interrogating beam, the beam of light shone onto the machine's housing, would have to be able to see a lot of machines and sequentially illuminate them. So we assumed that it would be positioned high above the sea of machines on the floor. Second, some of the interrogating beam would have to be reflected back to the source of the interrogating beam, where we would need a demodulator for the reflected light. Third, we would need a way to ensure that enough light was reflected from the machine's surface and sent to the demodulator.

Our no-holds-barred invention seemed to be a good and viable solution, providing that our assumptions were realizable, so we set about working on them for enablement of the invention. Getting the interrogator beam up high to see a host of machines was not a problem, and with a suitable control mechanism, the beam could be made to sequentially illuminate the individual machines.

An interrogator laser beam and reflected beam demodulator initially appeared to require a lot of work when we fortuitously learned that such a device had been developed. It was called a *vibrometer* and was commercially for sale. Things were looking up.

But how would we ensure that we got enough light sent back to the vibrometer demodulator from the light scattered off of the machine's housing? Well, we could attach a carefully aligned mirror to the machine's housing, but this would be tedious and a slight disorientation would send the reflected beam off target.

Then we thought of a corner reflector. This is a device that has three reflecting surfaces at right angles to each other and joined at a common vertex. Incoming electromagnetic radiation such as light bounces off one surface onto a second surface and then onto a third surface and is reflected back closely parallel to the direction by which it came. This would mean that slight disorientations would no longer be a problem, but we would have to purchase and attach a corner reflector to each machine. Then we learned of a commercially available retro-reflective tape that "may be easily affixed to a machine and allows a beam to be reflected at the same or similar angle of incidence as the incoming beam." And with that piece in hand, we had our invention. It is covered in U.S. Patent 7667827, and one of the patent's figures is shown in Figure 5-9. The parts table is shown in Table 5-3.

The POP Score: A Measure of Invention

Consider any patent. How far does it rise above the plane of practiced combinations? How big was the Eureka! for a patent? In an attempt to answer this, two of us suggested an heuristic of utility for studying combination inventions

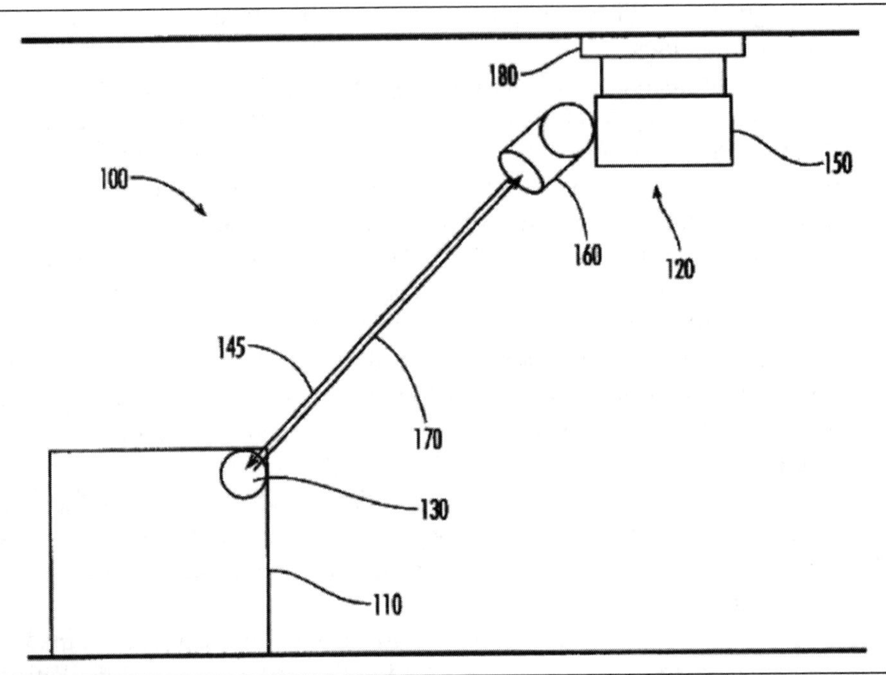

FIGURE 5-9 Figure 1a of U.S. Patent 7667827

Component number	Description
100	the remote vibration monitoring system
110	a remotely monitored machine
120	a vibration detection unit
130	a reflective patch
145	the interrogator beam (forward)
150	the control module (controls beam pointing direction)
160	the optics module
170	the vibration modulated beam (return)
180	an optional vibration isolation unit

TABLE 5-3 Patent 7667827 Parts

("White Space Patenting," Hershey and Thompson, *Intellectual Property Today*, Vol. 11, No. 8, August 2004). We called it the "POP score," which stands for Prescience of the Patent. The POP score formula is based on an averaging technique known as the *geometric mean*. It might look complicated, but it's really easy to do on a calculator—just plug in the numbers.

The POP score is defined with respect to a number of terms that you believe to be indicative of a combination invention. First you choose a time period for examination. This time period can be a specific year or a span of years. Then let

- T be the number of terms you want to examine
- Ni be the number of patents issued in the time period of interest that have term i in the fields of interest
- S be the number of patents issued in the time period of interest that have all T terms in the fields of interest

Then calculate:

$$\mathrm{POP} = \log_{10} \frac{(N_1 \cdot N_2 \cdots N_T)^{1/T}}{S}$$

Interpretation of the POP score is of course best done by the user for his or her particular technological area. As a start, however, the following information might be of help:

- For a POP score between 0 and 1, the combination of technologies is mature. The combination has been well noted by those of ordinary skill in the art, much time and effort has been applied by many different organizations, and efforts might already be well underway to create standards and interfaces.
- For a POP score between 1 and 2, there is definite interest and progress in combining the technologies. The elevator has left the ground floor but it might not be too late for an entrepreneurial effort to grow a valuable portfolio.
- A POP score of 2 or higher evidences innovation and breakthrough. To play in this area is exciting, and the opportunity to seize broad claims in IP is at its peak. Speed is of the essence.

Let's go back and calculate the POP score and take another look at the combination of the words "Internet" and "digital camera." Table 5-4 highlights the rapid decline in the POP score once the combination took off.

The Technology Linkage Diagram

The POP score looks into the past. It studies the progress of history already made. It's helpful and instructive to see from where we've come and how fast we've moved; but how do we seize opportunity and garner white space before it's colored?

Filing year	Internet	Digital camera	Internet and digital camera	POP
2002	17861	2620	147	1.67
2001	14978	2039	120	1.66
2000	11081	1463	67	1.78
1999	7031	950	37	1.84
1998	4225	621	21	1.89
1997	2297	337	10	1.94
1996	874	166	2	2.28
1995	199	84	1	2.11
1994	47	47	–	–
1993	23	35	–	–
1992	14	29	–	–
1991	8	23	–	–
1990	8	18	–	–

TABLE 5-4 The Progress of the POP Score for the Combination Invention

An Example of a POP Score Calculation

Just to show the arithmetic, let's calculate the POP score in Table 5-4 for the year 2000. The number of Internet patents was 11081, so $N_1 = 11081$. The number of digital camera patents was 1463, so $N_2 = 1463$. The number of patents having both terms is 67, so $S = 67$. There are two terms, so $T = 2$. We take the *T*-th root, which is then the square root, and we have

$$\text{POP} = \log_{10} \frac{\sqrt{11081 \cdot 1463}}{67} = \log_{10} \frac{4026.35}{67} = \log_{10} 60.09 = 1.78$$

One tool we can use is the *technology linkage diagram* ("White Space Patenting," John E. Hershey and John F. Thompson, *Intellectual Property Today*, Vol. 11, No. 8. August 2004). We pulse technologists for a set of terms denoting components that might be linkable or combined to form a useful system.

For example, suppose we're interested in *wireless remote sensing*. The following terms naturally come to mind:

- Sensor
- Location
- Encryption
- Display
- Cell phone

But now we have the question of what data sources to use to construct our technology linkage diagram. One source is the patent database, but it usually takes years before a patent is granted. Even patent applications publish 18 months after filing, so this data is quite old. Another source of data is the formal archival publications, but these are subject to lengthy review and publishing queuing delays. They might beat the patent data sources, but this data is also old. Sources that are quick to see print are trade journals or, as they are sometimes disdained, the "glossy rags." Whatever, they are rapidly published and can serve as a great insight to the future technology landscape. They can also serve to estimate the future patent landscape.

Table 5-5 shows the number of hits on the five terms and their ten possible pairs as found in a technical data abstract service in early July 2002.

We convert the data of this table into the technology linkage diagram shown in Figure 5-10 by representing the individual terms as circles whose radii are somewhat proportional to the frequency of occurrence or hits of the term. Links are then drawn between the circles. The link widths are somewhat proportional to the number of binary pairs normalized by a factor reflecting intersection data between the two terms.

Notice that the diagram includes a missing link. It is a good first bet that this might represent room for invention—and, indeed, it turned out that such was the case when we checked a year and a half later, as shown in Table 5-6. Note that the missing link has disappeared.

It is also useful to study the technology linkage data based on issued patent data. As we said, this data is old, but we can learn a bit about technology trend

Term	Hits
Sensor	30,411
Location	25,369
Encryption	955
Display	10,544
Cell phone	62
Term Pairs	**Hits**
Sensor and location	2179
Sensor and encryption	8
Sensor and display	731
Sensor and cell phone	2
Location and encryption	23
Location and display	644
Location and cell phone	6
Encryption and display	7
Encryption and cell phone	0
Display and cell phone	2

TABLE 5-5 Frequency of Occurrence and Their Pairs

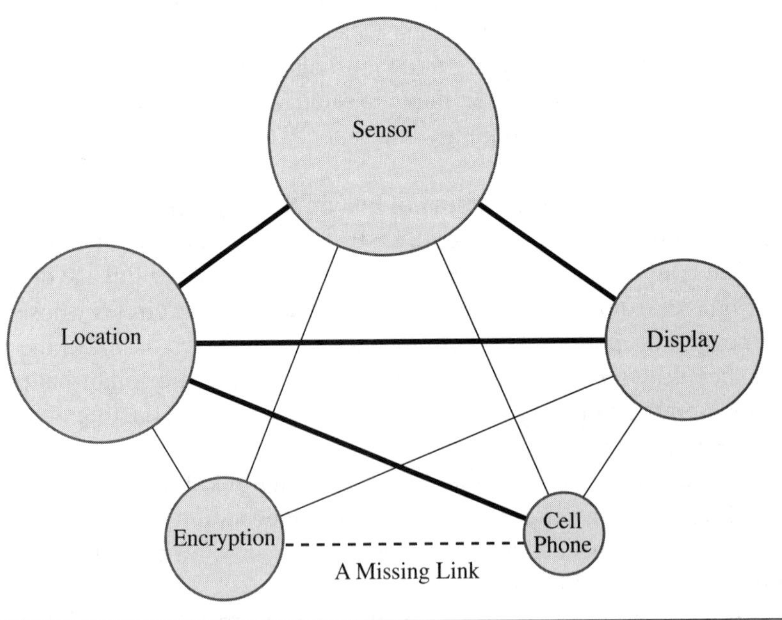

FIGURE 5-10 Technology linkage diagram

Term	June 2002 Hits		December 2003 Hits	
Sensor	76,790		92,236	
Location	99,442		115,863	
Encryption	2077		2475	
Display	77,894		93,706	
Cell phone	43		103	
Term pair	**June 2002 Hits**	**POP**	**December 2003 Hits**	**POP**
Sensor and location	8728	1.00	10,580	0.99
Sensor and encryption	51	2.39	72	2.32
Sensor and display	7088	1.04	8660	1.03
Sensor and cell phone	2	2.96	5	2.79
Location and encryption	333	1.64	393	1.63
Location and display	9999	0.95	11,903	0.94
Location and cell phone	15	2.14	30	2.06
Encryption and display	258	1.69	304	1.70
Encryption and cell phone	0	–	2	2.40
Display and cell phone	14	2.12	25	2.09

TABLE 5-6 Technology Linkage Data Based on Issued Patents

rates. Table 5-6 shows the counts of U.S.-issued patents between June 27, 2002, and December 7, 2003, whose patent claims contained the terms and term pairs in their patent claims. The POP scores are also computed. Note that the POP scores tend to decline with time. Lower POP scores tend to decline more slowly than larger ones, and this is to be expected as the technology of the system having the parts in question matures.

Barriers Dissolve with Time

In studying the POP scores of different combination inventions, it soon becomes clear that the crumbling of technological barriers is quite typical in an explosive technical field. We can sometimes expect this when useful elements come under performance pressure. The response is to preserve these elements by forcing invention to improve their performance and retain their utility. For example, the platform-neutral Java language grew steadily in prominence and importance, but in the late 1990s, it was sometimes assailed as being too slow or requiring too much memory.

These objections centered on issues that responded well to inexorable technological advances. Thus, it seemed natural that as computer enabling technologies such as memories and processors got larger and faster and yet remained physically small and required only a reasonably-sized energy source, we should have expected that Java would find its way into the cellular world. And, indeed, on March 16, 1999, *Computerworld* reported the following:

> *Sun Microsystems Inc. and the cellular phone unit of Nippon Telephone & Telegraph Corp. (NTT) said today they will begin studying ways to use Java and related technologies in future cellular phones and services to be offered by the Japanese carrier.*

Well, then, could a new technology emerge comprising mobile communications, platform neutral computation, and location technology? Was such a product on the way? With a little patent database research for the following simultaneous occurrence of the keywords in the claims, titles, and abstracts of the U.S. and worldwide patents as of August 24, 2001, it appeared so:

- Java
- Cellular
- GPS

The counts are displayed in Table 5-7.

The three dimensions of Java, Cellular, and GPS seemed to be a basis on which an exciting inventive thrust was being built. A beautiful example of a patent that hit on all three is U.S. Patent 6240360, "Computer System for Identifying Local Resources," issued May 29, 2001. The patent abstract keyed to the invention's third drawing (Figure 5-11), reads as follows:

Terms	Occurrences
Java	1103
Cellular	30,756
GPS	4573
Java and cellular	33
Java and GPS	13
Cellular and GPS	438
Java and cellular and GPS	7

TABLE 5-7 Counts of the Three Keywords in U.S. Patents

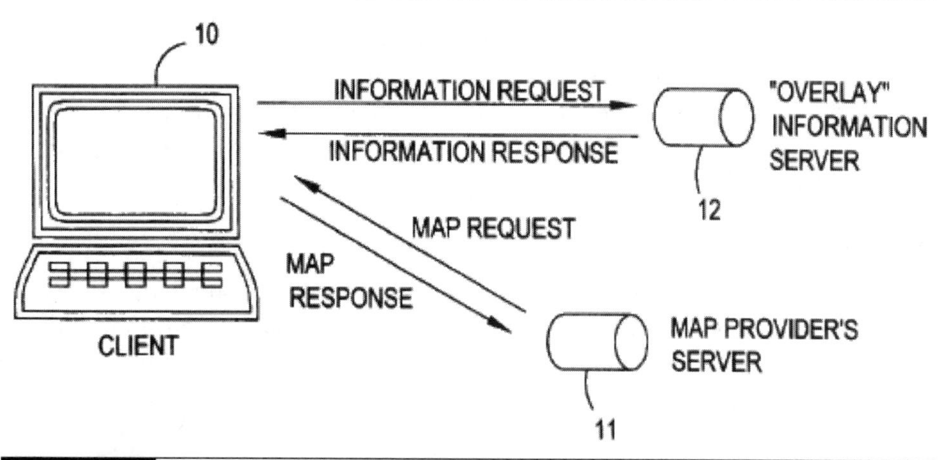

FIGURE 5-11 The third drawing in U.S. Patent 6240360

A map of the area of a client computer (10) is requested from a map server (11). Information relating to a place of interest is requested from an information server (12) by the client computer (10). The information is superimposed or overlaid on a map image at a position on the map image corresponding to the location of the place of interest on the map. The information (or "overlay") server (12) may contain details of, for example, hotels, restaurants, shops, or the like, associated with the geographical coordinates of each location. The map server (11) contains map data, including coordinate data representing the spatial coordinates of at least one point on the area represented by the map.

The coming together of the three dimensions is reflected within the embodiments:

- *In a preferred embodiment, a client device which has the capabilities of both a cellular telephone and a Web browser may pass the names and/or geographical coordinates of its surrounding cellular base stations to the map and/or overlay server computers.*
- *In a particularly preferred embodiment, the client computer may include locating means for establishing the current geographical location of the client computer. This may be by means of a satellite system such as the Global Positioning System.*

- *In a more advanced embodiment, suitable for client browsers capable of running Java or some other local processing capability, the response from the information server 12 specifies one or more overlay icons and associates a longitude and latitude with each. Longitude and latitude are resolved to screen position by a Java Applet or other locally executed program.*

Recap

In this chapter, we have used our insight to dimensions to motivate Eureka! moments to produce and study combination inventions. We have introduced tools and outlined approaches to inventing as well as a method for measuring the Eureka! moment.

Discussions and Reflections

- Consider one of your company's maintained patents that is midway through its life. Try to identify what the preferred embodiment was at time of filing. Is it still the preferred embodiment?
- How do the inventors in your company stay sensitized to inchoate technologies that could be used to fulfill critical functions in their inventions? How are claims and the specification prepared so that associated reasonable contingencies are covered?
- Most inventions, particularly those from the social inventors—the majority of the inventors at an R&D facility, for example—are of the improvement variety. They improve upon an already existing product. They are advances, to be sure, but most of them are one-dimensional. The combination inventions, however, are multidimensional, with the synergy of very different technologies. What percentage mix between improvement and combination invention makes sense for your company's R&D effort? What percentage of your company's R&D inventors are capable of true combination invention?
- Review the products manufactured by your company. Are there any as yet unrealized combinations that could result in a new product of utility to an emerging social trend, new regulatory environment, or new technology? If so, attempt to compute the POP.
- Expand your circle of thought outside your company's boundaries. Do you know of any firms with whom a meaningful combination invention could be forged under a joint venture?

- Try an "innovation circle." In a group of players, each one thinks silently of an emerging technology and writes it on a card. When everyone is ready, the players post their cards on a common wall. The group then considers how the technologies might be combined to produce a combination advance.

- Dependent claims revisited: In reality, there is often no bright line between improvement and combination invention. They are at best useful but rough characterizations. In Appendix A, we say that generally a series of dependent claims narrows the invention as a series of dependent claims covers ever less of the space of the independent claim. In the Appendix, we also say that a claim dependent upon claim 1 must have its entire boundary within the claim from which it depends and its total area must be smaller than the claim from which it depends. This is true, but now that you have studied dimensionality, you can see how a dependent claim can reach into other dimensions and transform what was an improvement invention into one that is a combination invention.

  As an example, take a look at U.S. Patent 5758313, "Method and Apparatus for Tracking Vehicle Location." This patent was directed toward computer aided dispatching for fleet management. In this business, it is essential that vehicle location is known, and how to efficiently operate a fleet of vehicles is paramount. Map displays and overlays are useful tools.

  Let's look at Claim 1 in the patent:

  1. A computer aided dispatching method comprising:

  providing a display comprising a first display segment and a second display segment, said first display segment comprising a digitized representation of a raster map and a plurality of user locatable marks, each of said plurality of user locatable marks representative of one of a plurality of mobile units at a mobile unit position, said second display segment comprising vector text information for at least one of said plurality of mobile units; and

  using a computer aided dispatch system operably coupled to said display; said computer aided dispatch system comprising order data from customers, said order data having a portion being transferred from a data acquisition device coupled to said computer aided dispatch system to a radio in one of said plurality of mobile units.

  Claim 1 is an independent claim. The patent could have issued with just this one claim for its useful, novel, and nonobvious invention. It is primarily in the dimensional space of graphical displays. But now look at dependent claim 6:

> *6. The method of claim 1 wherein each of said plurality of mobile units comprises a navigation tracking device, said navigational tracking device comprising a microprocessor operably coupled to a global positioning system (GPS) navigational sensor and a mobile radio modem operably coupled to said microprocessor.*

With this dependent claim, the invention has reached and coupled itself into the other dimensional spaces of location-determining technology and wireless communications. With this dependent claim, an improvement invention has become a combination invention. What other examples come to mind?

- The biggest job in securing white space is in forecasting technological advances, societal changes, and anticipating regulation. This is where a grounding in science or technology is very important. For example, combining a vibration sensor, an RF transmitter, and a glue gun is not innately compelling. (Substitute a motor for the glue gun and we sense a useful system for the present.) Thus, we expect there is ample white space available for the glue gun, vibration sensor, and transmitter system, but the value of patents rising out of this space would, most likely, be nominal at best. But, in time, perhaps robotic systems will be gluing parts together and then what would be the value of such a patent?
- The Markey & Antheil patent on the frequency hopping device gamed the system of jamming a transmission by moving the transmission frequency around in a pseudo-random manner. This technique is an electronic countermeasure. A counter-countermeasure is to find the new transmission frequency and move the jamming signal to the new transmission frequency before a significant amount of data is transmitted. In a sense, the system of operation is such that the new transmission frequency attracts the jammer. Is there a way that this system could be gamed by adding another pseudo-random mode that occasionally communicates data by not sending data directly but rather by the selection of the frequency of the transmission itself? In this mode, the jammer would actually aid the transmission of data.

Part II

Seeking a Eureka! Moment

Chapter 6

Law, Regulation, and Standards

The people must and will be served.

—William Penn, *Some Fruits of Solitude*

Laws, regulations, and standards are external motivators of Eureka! moments affecting public safety, security, and interoperability. This chapter explores these subjects with a view toward the opportunities for an inventor. Laws and regulations both feed on inventive activity, and standards are the crystals that emerge from these oscillations. Standards are produced from the most mature of the developed thoughts, and they require specific and targeted invention for their practice.

Regulation comes to us from many sources; the most obvious situation prompting regulation occurs when an object or situation poses a clear and present danger and requires immediate control. But more subtly, a technology can arise that enables all sorts of safety benefits to the public. This then gradually motivates adoption and finally regulation. If you can identify an emerging pre-regulation environment and innovate so that you seal up the white space requisite to cost-effective practice under the eventual regulation, then you have grabbed the brass ring.

Safety Regulations and Invention

We've looked at the growing movement to use light emitting diodes (LEDs) in traffic signals and reviewed some of the main reasons for doing this, including the LED arrays' lower power demands, the greater brightness of the LED arrays, and the lower maintenance required for them. There is, however, one more very interesting benefit to using LED arrays instead of an incandescent bulb: LED signals can be significantly *safer* than incandescent signals. The reason results from electrophysics: as shown in Figure 6-1, LEDs turn on faster than incandescent light sources and, therefore, alert drivers in a more timely fashion ("New ITE Standards for Traffic Signal Lights," C. Andersen, FHA/DOT, December 14, 2005).

If used in car signal lights, this decrease in signal alert time translates directly into shorter reaction times and stopping distances and fewer rear-end accidents. The benefits to traffic safety are profound. The European Lamp Companies Federation reported that LEDs "illuminate 99% faster than incandescent car lamps resulting in a reduced stopping distance of 8 m at 100kph. Eight meters equals the length of up to 2 cars and can be the difference between life and death in a head-tail collision" (www.energy-efficiency-watch.org/fileadmin/eew_documents/Documents/Community/Elcfed/04DYK_5_LED_06.pdf).

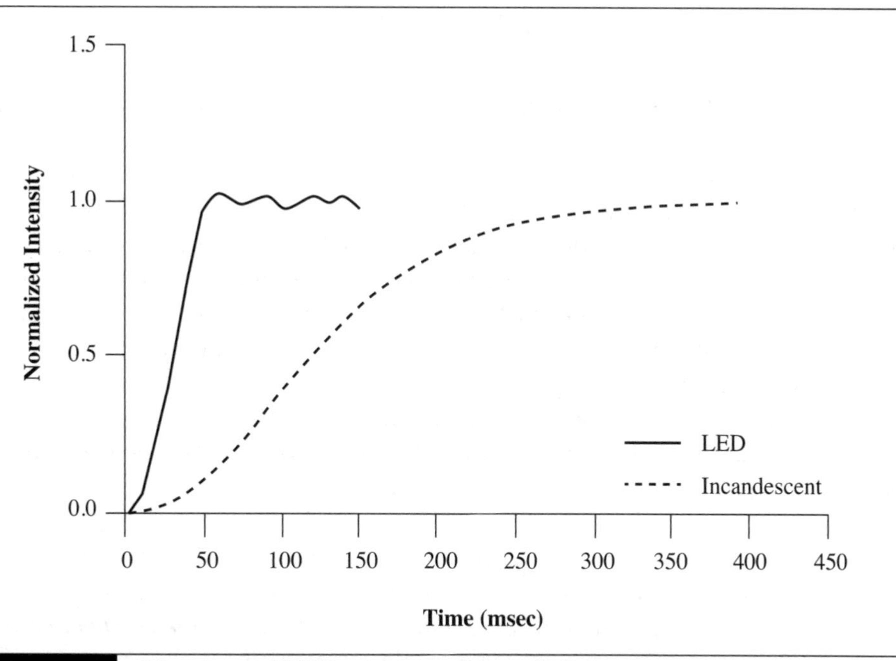

FIGURE 6-1 Rise times of LEDs and incandescent bulbs

This last point is extremely significant, because it intimately affects the commonweal and thereby becomes a candidate for regulatory consideration. The basic question is this: If you can significantly reduce the risk to the public, or better shield your customer or client from risk, and there exists the reasonably attainable technological means to do so, do you have an imperative to act?

T. J. Hooper Case

The *T. J. Hooper* Case (1931–32) considered such a question and provided case law impact that has rattled down through the decades. An operator of the tugboat *T. J. Hooper* was sued on the loss of two barges and cargo in a gale. The elements of the case follow:

- The barges were in transit off the coast of New Jersey.
- The charges were based on the tug's not having a reliable radio that would have facilitated reception of gale warning that had been timely broadcast.

- Other tugs that were so equipped had responded by putting themselves out of exposure.
- The *T. J. Hooper* was found unseaworthy because it lacked a reliable radio.

At that time, tugs were not required to have radio sets on board for this purpose, but the judge ruled that the "standard of seaworthiness changes with advancing knowledge, experience, and the changed appliances of navigation."

On appeal, the famous Judge Learned Hand, often called the tenth judge of the Supreme Court, affirmed and commented: "A whole calling may have unduly lagged in the adoption of new and available devices. It never may set its own tests, however persuasive be its usages."

A soul-searching article asked some important questions ("Legal Ethics in an Electronic Age: Where No One Has Gone Before?" *National Law Journal*, March 24, 1997):

> *Today, we communicate with clients by fax, cell phone and e-mail. The first two are laughably insecure. But e-mail can be encrypted, so if a message is misdirected or intercepted, it is possible for the privacy of the communication to be protected: the wrong recipient would see only gibberish.*
>
> *Encryption is relatively rare today. But software is available that will almost effortlessly encrypt an e-mail message. At what point does the failure to use encryption make a lawyer as liable for an accident as the captain of the T.J. Hooper? Next year? Next month?*

GPS

Technology can drive regulation, and vice versa; the two are intimately linked. A good example of an advance in technology, an important combination invention that arose out of forthcoming regulation, has its roots in the Global Positioning System (GPS). The GPS has been called a national treasure, equipping users with the ability to determine their global positions and to know the time, almost worldwide and at a reasonable and continually declining cost. In essence, new dimensions were provided for engineers to combine with other pieces of art to create new components and services of immense value and protection for the public. In retrospect, it was only a matter of time before regulation entered the picture, and it did so via the Federal Communications Commission's (FCC) declaration on Enhanced 911 (E911) service: "Wireless carriers are required to provide Automatic Location Identification (ALI) as part of Phase II E911 implementation beginning October 1, 2001."

A patent count, made in early 2000, showed that GPS-related patents were assigned to many companies and organizations, including the United States government. Qualcomm was well represented and a small number of patents were held by a company called SnapTrack. A significant technology play soon followed the FCC's position with an announcement in *GPS World*, March 2000, foreshadowing a reported billion-dollar acquisition:

> *Qualcomm plans to combine SnapTrack's patented Wireless Assisted GPS technology with Qualcomm's GPSOne system, a Mobile Station chipset and software combination for code division multiple access (CDMA) cellular and personal communications service (PCS) networks…to meet the Federal Communications Commission deadline next year for the capability of locating people placing emergency calls to public safety operators.*

In its broadest sense, industry is concerned with coloring white space and protecting it with patents. Invention is the most obvious route but not the only one. Coloring white space extends to acquiring patents that others have in order to augment your portfolio to meet a regulatory challenge. So always pay attention when you see the word *safety* in connection with technology. The reason is shown in the simple societal oscillator of Figure 6-2. If you invent, develop, or even demonstrate a technology that reduces the public risk, you might be helping to frame legislation, and this can be a double-edged sword.

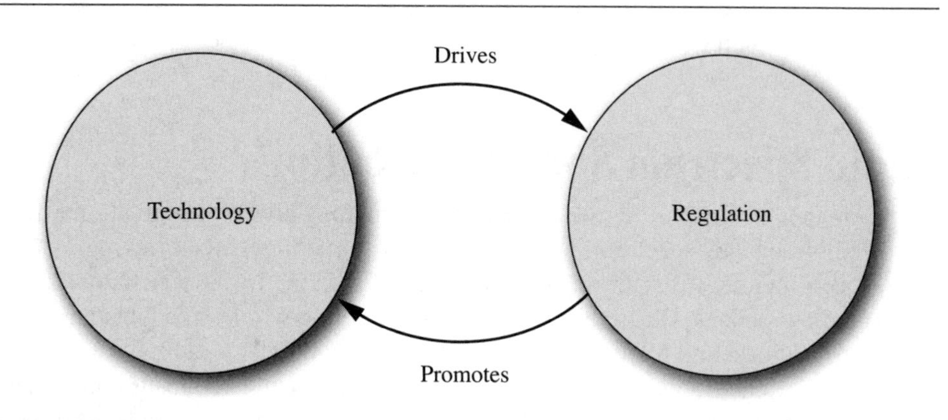

FIGURE 6-2 The technology-regulation oscillator

A Short, Personal Story

Let me tell you how the tie between technology and regulation came home to me many years ago. I had been temporarily assigned to a customer's team charged with suggesting ways in which electronic technology might be used in a better way to characterize hazardous waste sites. We were all eager to help save the planet. The ideas flowed. But then we had our first review. We reported our conclusion:

Electronics can be used to speed up the characterization process.

"Well," our reviewer looked slightly uncomfortable, "we're not really looking for an increase in speed." And back to the drawing board we went.

The time for our second review came and we had a new conclusion:

Electronics can be used to find toxic materials not yet on the list of chemicals we're concerned about.

The reviewer seemed to stiffen and suck the air through clenched teeth. "Not really what we're looking for."

The third review came and our third conclusion was ready:

Electronics can be used to detect those chemicals you are concerned about but at concentrations far smaller than the limits of concern.

"No. Not what we want."

This time we were bold. We asked, "Why?"

"Because" the reviewer surveyed the naive brain trust, "we believe that the government sets the acceptable limits based on the existing technology."

"Well, then," we were at our wit's end, "exactly what do you want?"

"Ah," the reviewer looked relieved, the ritual was playing itself out. "What we really want is electronics that will make it cheaper to characterize a waste site. Should we want to, of course."

Radio Spectrum Spur to Invention

Another example of regulation driving technology concerns regulation affecting radiofrequency spectrum. It proceeded from some farsighted thinking by the spectrum's stewards, the U.S. government. E-ZPass, fire alarms, baby monitors, cordless phones, Bluetooth, and a panoply of other wireless gadgets have mushroomed and filled catalogs and storefronts. Why?

They came about because people wanted them. Wireless means convenience—no muss, no fuss. All right, but how? How can I just transmit signals? Don't I have to apply to the government for permission? Pay a fee? File some forms? Get a license?

The wonderful answer is no. The government, in the early 1980s, did a proactive, pro-public promotion that energized the regulation-to-technology transition. In 1981 the Federal Communications Commission (FCC) adopted a Notice of Inquiry (the boldface is mine):

> *...for the authorization of certain types of wideband modulation systems. The Inquiry is unusual in the way that it deals with a new technology. In the past, the Commission has usually authorized new technologies only in response to petitions from industry. However in the case of spread spectrum, the Commission initiated the Inquiry on its own, since its current Rules implicitly ban such emissions in most cases, and* ***this prohibition may have discouraged research and development*** *of civilian spread spectrum systems. As the next step in this proceeding, we are proposing in this Notice of Proposed Rulemaking rules that would authorize the use of spread spectrum under conditions that prevent harmful interference to other authorized users of the spectrum.* ***We anticipate that this authorization will stimulate innovation in this technology****, while meeting our statutory goal of controlling interference. (FCC 84-169 98 F.C.C.2d 380)*

And stimulate innovation it did! One of the chunks of spectrum allocated by the FCC for these "unlicensed bands" is from 2400 to 2483 MHz. Look at Figure 6-3, which shows a spectrum survey done by the National Telecommunications

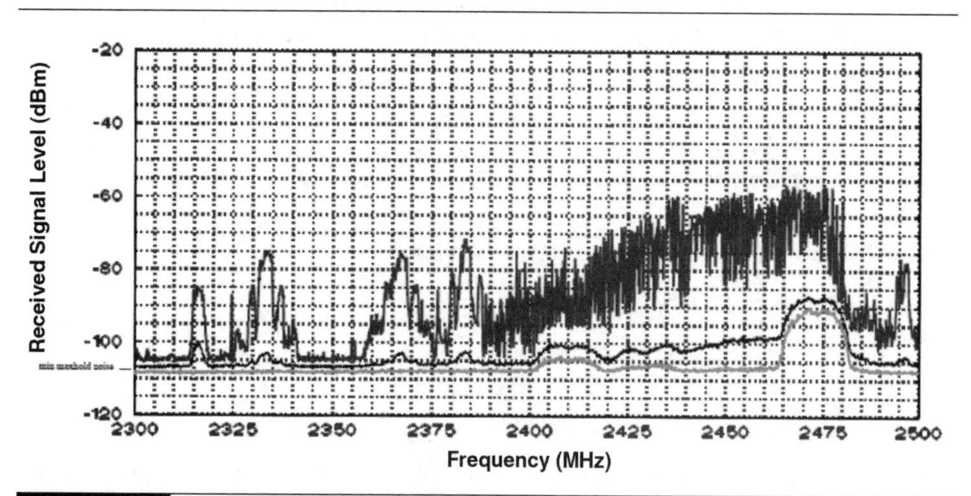

FIGURE 6-3 NTIA spectrum survey (NTIA Report 97-334 "Broadband Spectrum Survey at San Diego, California")

and Information Administration (NTIA). It summarizes 28,800 sweeps across the 2.3–2.5 GHz frequency spectrum taken in San Diego, California, in 1995 and demonstrates the crowding of wireless transmissions into the 2400–2483 MHz unlicensed band.

How Standards Can Stimulate Invention

A new technology can also beget a standard—one of engineering's greatest contributions as it promotes efficiency and interoperability. After all, it's nice to be able to plug a lamp or another appliance into the same socket as a toaster. What a standard setting organization does is work toward an industry agreement for a product to do the following (from American Intellectual Property Law Association, "Antitrust: Current Trends in Product Tying and Standard Setting Boards," 2007):

- *Increase the extent of product compatibility and interoperability;*
- *Increase the speed of product development and distribution;*
- *Ease the entry of market participants on a wider scale; and*
- *Advance the globalization of markets.*

Patent Pooling

Many different companies can contribute or "pool" their patents to create a product. The U.S. Department of Justice (DOJ) refers to this patent pooling as "an aggregation of patent rights for the purpose of joint package licensing" and is concerned that a proper position be reached between patent law and antitrust law, a tension between efficiency and monopoly.

A pro-competitive patent pool functions to do the following:

- Integrate complementary patent rights
- Reduce transaction costs
- Clear blocking positions
- Avoid costly litigation

The pool is filled with what are termed "essential" patents. Each of these patents should meet the following requirements:

- It should have no substitutes—the patent should not compete with any other patent whether within or without the pool
- Is needed to comply with the standard
- Is necessarily infringed on implementation of the Standard's specifications

The meaning of the last point in the list above is critical to understanding an essential patent. The term "infringement" means to practice an invention protected by a valid patent. The term "infringement" is pejorative if used in the sense of an unauthorized practice of someone else's valid patent, but in the sense that it is used here it means that in order to practice the standard it is required to practice the essential patent; i.e., there is no other way to perform the standard than to infringe a patent essential to the standard.

Essential Patents

Here's the DOJ reasoning concerning the definition of an essential patent, highlighting the antitrust concerns:

> *An inclusion criterion broader than "essentiality" carries with it two anticompetitive risks, both arising from the possibility that the pool might include patents that are substitutes for one another and not just complements. Consider, for example, a situation in which there are several patented methods for placing DVD-ROMs into packaging—each a useful complement to DVD-ROM manufacturing technology, but not essential to the standard. A DVD-ROM maker would need to license only one of them; they would be substitutes for each other. Inclusion in the pool of two or more such patents would risk turning the pool into a price-fixing mechanism. Inclusion in the pool of only one of the competing non-essential patents, which the pool would convey along with the essential patents, could in certain cases unreasonably foreclose the non-included competing patents from use by manufacturers; because the manufacturers would obtain a license to the one patent with the pool, they might choose not to license any of the competing patents, even if they otherwise would regard the competitive patents as superior. Limiting a pool to essential patents ensures that neither of these concerns will arise; rivalry is foreclosed neither among patents within the pool nor between patents in the pool and patents outside it.*

(continued)

The aspiring patent pool members hire an expert to be the umpire of essentiality, and this expert's decisions regarding essentiality are conclusive and may not be appealed. The expert is subject to the following restrictions that function to reduce any actual or perceived conflicts of interest:

- The expert is appointed and removed by a majority of the licensors.
- The expert's compensation is based on time evaluating patents and not a function of the essentiality decisions.
- The expert will not be economically affiliated with any individual licensor.

The division of royalties is another very important agreement. Typically, the division of royalties to a licensor is according to the following criteria:

- The number of a licensor's essential patents
- How often a licensor's essential patents are infringed
- Age of essential patents (newer weighted more heavily)
- Whether essential patents relate to optional or mandatory features

If you invent and patent what is deemed to be an essential patent for a standard, you might have grasped a gift that keeps on giving. If this is your goal, keep your eye on what future product is coalescing and what niche you might be able to fulfill with, most likely, a small invention that has no alternative but is compelling in its promise to make the product possible in some dimension.

Standards and Cryptography

A standard need not be produced by a group effort. Sometimes a single entity, a person or a company, will create something that is so useful that it becomes a standard for the government or for a commercial movement itself. Such has been the case with cryptography, the art turned science for concealing the content of communications or of proving identity or intention through a means of message authentication.

Cryptographic Protection—A Reluctant Path

The desire to protect one's thoughts is as old as civilization itself. Julius Caesar employed a cipher system that bears his name, and through the ages the outcomes of countless battles have been tilted one way or the other by cryptographic successes and cryptographic blunders.

What is at first surprising to most folks is that cryptographic methods were not adopted more widely much sooner for communication that travels through an electronic communications network. It would seem to be no more than basic communications hygiene to deploy and support such methods in light of the threat environment that is so ubiquitous.

The reasons for the slow start are at least three. First, early cryptographic systems cost the users in currency as well as via a negative impact on operational efficiency. Second, initially cryptographic prophylaxis didn't fit well with our culture of openness and, even worse, we couldn't measure it in terms that made visceral sense to us—no miles per gallon, no more bang for the buck, nothing except a vague promise of reducing the "threat." But, and this was the third problem, what was this "threat"? Show me the threat and I'll show you the money. The problem with this pushback was that if the threat were really top notch, you wouldn't be aware of it. So cryptography remained a black art, practiced behind closed doors in the government and guarded by those who do or die.

Communication used to be simple, more black and white: a piece of information was either classified or it wasn't, with no well-recognized middle ground. But in the 1970s things were changing, and by the late '70s President Carter had concluded that the country's national telecommunications protection policy should comprise a number of elements, including the following enumerated in PD-24 (Presidential Directive 24; see http://www.fas.org/irp/offdocs/pd/pd24.pdf):

- *Government classified information relating to national defense and foreign relations shall be transmitted only by secure means.*
- *Unclassified information transmitted by and between Government agencies and contractors that would be useful to an adversary should be protected.*

(continued)

- *Non-governmental information that would be useful to an adversary shall be identified and the private sector informed of the problem and encouraged to take appropriate measures.*

This was indeed high-level recognition that such information as, for example, an unreleased decision for a federal funds rate target by the Federal Open Market Committee might deserve cryptographic protection as well as plans for a weapons system that might never be built, and, if it were, might not ever work.

One-Key Systems

PD-24 motivated the technical moves toward protection of unclassified but non-public information in the 1970s as the government became concerned about protecting data on U.S. citizens such as Social Security and personal data that it had lawfully collected. The National Security Agency (NSA), an arm of the Department of Defense charged with providing encryption equipment to the government, was less than eager to try to develop a cryptography that could be used to protect this nontraditional class of information, and it eventually fell to the nation's metrologists, the then National Bureau of Standards (NBS), to come up with a proper cryptographic system. From their efforts emerged the Data Encryption Standard (DES) and later the Advanced Encryption Standard (AES).

Harvesting the Public's Intellectual Capital

The NBS decided to leverage the system to come up with a new encryption scheme, and their system was the great and wide public of the nation. They advertised in the *Federal Register* for an appropriate technique with which to fulfill their new mission. And the public didn't let them down, although it did require more than one solicitation.

The best invention to show up was an algorithm from IBM. The NBS grabbed it, scrubbed it, advertised it, and eventually promulgated it as the Data Encryption Standard. It became a huge success and was widely employed for a variety of tasks. It continues today as the cryptographic workhorse in many installations, although it is being aggressively replaced by yet another standard, the Advanced Encryption Standard.

The DES and AES belong to the class of cryptographic techniques known as *one-key* or *symmetric key* systems. In such a system, the sender and receiver must have the same cryptographic keying variable to communicate securely. The keying variable is a set of random characters that is entered into a cryptographic device and governs the way it operates. It is a tenet of modern cryptography that the algorithm or cryptographic principle is assumed to be publicly known, and therefore the security afforded by the cryptographic system does not rest in hidden knowledge of the system but rather in a secret quantity—that is, the keying variable that is entered into the cryptographic system just prior to its operation. There's a good reason for this basic rule, as it would be, unfortunately, just a matter of time before an adversary acquired a secret cryptographic system. There are losses, robberies, and people who will trade in almost anything for money.

So, for two correspondents using the DES or the AES to protect their communications, both correspondents must be in initial possession of the same secret keying variable and must enter it in their cipher equipments before communicating securely. Such a requirement imposes a difficulty on the correspondents, as they must somehow come into possession of the same keying variable. This might require a trusted courier. It certainly will require a delay if both start from scratch. It's an even bigger problem if they want to communicate securely with several parties.

What could get around the problem of having to distribute identical keying variables to all correspondents prior to communications? What was needed was a way for two parties to create a way of communicating securely without this intensive distribution problem. But how could it be possible to arrive at a

Unanswered Requirements

I believe that what eventually gave grief to the NSA was the challenge to satisfy a government customer base requiring something more than two-party secure communication links. For years, the DOD had been in the unchallenged driver's seat when it came to cryptographic policy. It was a position that seemed to say, You want to communicate securely? Then you must use our equipment the way we say to use it. But the world of communications was spinning ever faster, and networks were growing into complex networks with many users attached and many levels of communications protocol. The old two-party solutions with their symmetric key cryptography just couldn't cut it anymore. The dog had simply grown too large for the tail to wag it.

commonly held secret when *all* of your communications were visible to any interested viewer? It seemed a problem that should have been easily proven impossible—but it wasn't.

Public Key Cryptography

The first "thaw" came when a student discovered a clever, but impractical, way for two parties, using completely open communications, to develop a common secret with less work than would be required of any observer (interceptor) to find the common secret so developed, and the flood of innovation and invention flowed. Public key cryptography (PKC) developed quickly. It is a very important innovation, because it allows the establishment of secure communications without first distributing a secret keying variable to the communicating parties.

Finally, one particular system had become widely used and seemed to hold up well under the withering fire from the college profs. This was the RSA, a system named after its creators (Ron Rivest, Adi Shamir, and Leonard Adleman). The only problem with the system's implementation was the size of the numbers that had to be used to run it to render the latest attacks impotent. These mammoth integers were causing all sorts of problems, from lengthened communication times to uncomfortable computation times.

PKC Underpinnings

Mathematicians had become convinced that operations that had inverses, such as multiplication and division, were sometimes more easily accomplished in one direction than the other. In particular, *exponentiation*, taking a quantity to a power or exponent, was relatively easily performed as opposed to the inverse problem, that of finding the exponent given the *exponentiated* quantity when these operations were performed in what is called a *finite field*. That was a long sentence, but you can think of it in the following way: If I form $z = x^y$, this is exponentiation. I have raised x to the y-th power to form z, and in a finite field this is done relatively easily. If I want to find y given z and x, this is the inverse operation, and in a finite field this is relatively difficult. What PKC does is to structure a technique that requires two correspondents to do two relatively easy exponentiations to develop a mutually held quantity. For someone else, e.g. an observer or interceptor, determining this quantity is a hard inverse problem.

This asymmetric difficulty is at the heart of most practical public key cryptography. (It is sometimes called two-key or asymmetric cryptography because the correspondents in the preceding paragraph are choosing their exponents randomly and therefore differently.) The PKC arts have grown over the years, but one nagging problem remains: The security of a PKC system is a moving target—that is, you can often show that breaking (the old word was "cracking") a specific PKC system is *no more* difficult than solving a particular problem in mathematics, such as factoring a large composite number. What this means is that the difficulty of breaking a particular PKC system **is not greater** than solving a known problem. (The mathematicians call this *upper bounding*.) But how far below the upper bound, no one knows. And so, over the years, mathematicians have continued to discover better, faster ways to break a particular PKC system.

(As an aside, this has been a great boon for mathematicians. I'll never forget how thrilled a prominent mathematician was at a conference when he announced that he had received a grant of $15,000 to work on factoring numbers. Problems they loved were suddenly in style—and there was cash.)

Elliptic-Curve Cryptography

There then entered a white knight: a new system, elliptic curve cryptography, was born and provided a security comparable to the RSA against the most potent attack but with numbers, and concomitantly reduced computational times, of a fraction of the RSA's requirement. This was a great development, and today you will find elliptic curve cryptography used in many communications architectures, even extending into the standards realm. The founders of the new cryptographic system applied for and were granted a suite of patents covering their invention.

And now for the best part. The government was so convinced by the new system that it purchased the patents for many millions of dollars. This was a reversal of role, a celebrated triumph of an industry starting from scratch and rising to recognized and rewarded prominence.

An article in eWeek.com ("NSA Buys License for Certicom's Technology," Oct. 24, 2003) included the following:

> *In an extraordinary move, the National Security Agency has purchased a license for Certicom Corp.'s elliptic curve cryptography (ECC) system, and plans to make the technology a standard means of securing classified communications.*

As part of the $25 million agreement, the NSA can grant sublicenses within a limited field of use. This most likely will include other government agencies, federal contractors and other parties that send sensitive data to the agency.

This is the first time that the NSA has endorsed any sort of public-key cryptography system.

A Eureka! Moment: Turning a Problem Into an Aid Public key cryptography is advantageous because it allows two parties to establish a mutual secret "over the air" without any prior collaboration, and, as we said, most PKC systems are based on finite field exponentiation. But is there another way? A way that exploits physics rather than mathematics? Some colleagues and I came up with such a different approach if the parties are in a busy urban environment. Our Eureka! moment was motivated, and indeed made possible, by a communications problem.

Recall the discussion about improving a microwave oven in Chapter 2. We appealed to the analogy of occasional poor reception, or "cold spots," resulting from the multiple reflections of the microwave signal bouncing off moving reflectors. We realized that we could exploit this problem phenomenon to establish a secret keying variable between two users. Our invention also made use of the radio communications principle of reciprocity. This principle shows that the characteristics of the radio propagation path between two parties are the same for transmissions in either direction. The channel changes very rapidly due to the many moving reflectors and hence the reciprocity is short-term. At very high radio frequencies, the path characteristics are quite random in time and also in position.

We used these behaviors to generate keying variables, where in essence the communications path becomes the variable. We named this invention "channel-based cryptography," and it resulted in U.S. Patent 5604806, "Apparatus and method for secure radio communication." The abstract sums it up:

Characteristics of the radio channel are used to establish key sequences for use in encrypting communicated information. These characteristics are the short-term reciprocity and rapid spatial decorrelation of phase of the radio channel. The keys can be established with computations equivalent to a bounded distance decoding procedure, and the decoder used to establish a key may be used for processing the subsequent data transmission. Compared to classical and public-key systems, an alternative mechanism for establishing and sharing key sequences that depends on a physical process is provided in which each party need not generate a pseudorandom quantity because the necessary randomness is provided by the temporal and spatial non-stationarity of the communication channel itself. By using a channel decoder, the probability of two users establishing the same secret key is substantially unity, and the probability of an eavesdropper establishing the same key is substantially zero. Also, the number of possible keys is large enough that finding the correct one by exhaustive search is impractical.

Today cryptographic inventions abound, and much white space remains for your Eureka! moments. As in other cases, analogy can be a powerful way to engender a Eureka! moment—and, don't forget, a problem in one field can sometimes be turned to a significant advantage in another.

Recap

This chapter has attempted to show the link between the governmental processes of law and regulation with the significant opportunities for invention and the standards that proceed from a state of mature invention. The last subject considered was that of cryptography. Cryptography was included because it is a technology that has become crucial to the effective and efficient functioning of today's commerce and industrial operations. It has its own set of problems that will require our best innovative efforts for years to come. One of these problems is the man-in-middle-attack introduced in Chapter 3. This attack is particularly troublesome for many instantiations of public key cryptography, and we explore this in one of the problems in the following "Discussions and Reflections" section.

Discussions and Reflections

- Has regulation or deregulation had any impact on your company's patenting focus?
- Might any of your company's inventive activity make a positive safety impact on the industry or art?
- Inventing is never over. In fact it's quite often an unending chain of Aha! –Oh! –Aha! –Oh! … Let me try to convince you by returning to the LED traffic light. The LEDs seem to be ideal replacements for incandescent bulbs, after all they cost less, require less maintenance, and appear to make the traffic scene safer. And they most likely do these good things, but do they come at a penalty in another dimension of safety? Take a look at the following (from http://www.msnbc.msn.com/id/34436730/):

> *MILWAUKEE Cities around the country that have installed energy-efficient traffic lights are discovering a hazardous downside: The bulbs don't burn hot enough to melt snow and can become crusted over in a storm—a problem blamed for dozens of accidents and at least one death.*
>
> ...
>
> *Many communities have switched to LED bulbs in their traffic lights because they use 90 percent less energy than the old incandescent variety, last far longer and save money. Their great advantage is also their drawback: They do not waste energy by producing heat.*
>
> ...
>
> *Illinois authorities said that during a storm in April, 34-year-old Lisa Richter could see she had a green light and began making a left turn. A driver coming from the opposite direction did not realize the stoplight was obscured by snow and plowed into Richter's vehicle, killing her.*
>
> ...
>
> *Authorities said dozens of similar collisions have been reported in other cold-weather states, including Iowa and Minnesota.*

Does this sort of problem call for invention? It sure does and, for an example, look at a patent application filed on November 18, 2005, with the USPTO that matured into U.S. Patent 7211771, "De-icing system for traffic signals," with abstract:

> *"A circuit is disclosed for detecting and eliminating the buildup of snow and/or ice on the viewable face of an LED traffic signal lens. The circuit measures the ambient temperature within the LED signal, and when the temperature falls to a level where snow and/or ice accumulation can occur, the circuit begins looking for snow and/or ice buildup on the lens of the LED signal. An infrared LED transmits a signal which is reflected when snow or ice is present on the lens of the traffic signal. When the reflected signal is received by an infrared receiver, it sends a signal to a microcontroller, which analyzes the signal to determine if it is a valid signal. If it is, a heater is turned on until the ice and snow are removed."*

- Can you think of any other cases in which an improvement might create a safety concern? How about electric vehicles? They are much quieter than conventional vehicles. Might people with impaired hearing be more endangered by them? As many motorcyclists say, "Loud pipes save lives." What do you suggest to address the electric vehicle quiet operation concern?

- The snow and ice problem with the LED traffic signals is an example of a pitfall awaiting an improvement invention that uses a new technology. But sometimes the societal danger comes from the public's adoption of and addiction to the new technology. Consider texting for example. Figure 6-4 shows the explosive growth of SMS (Short Message Service) in the United States (from http://en.wikipedia.org/wiki/SMS).

  Obviously, the technology has become integral to society's form and function; obviously, the potential for dangerous distraction has rapidly escalated. We all know of the hazards of texting while driving and even walking. What is surprising to me, however, is the seeming lack of foresight by regulators and enforcement authorities in not getting ahead of the problem. A case in point is seen in the Government Accounting Office's (GAO's) study on the effectiveness of the Federal Railroad Administration's (FRA's) efforts to reduce the train accident rate

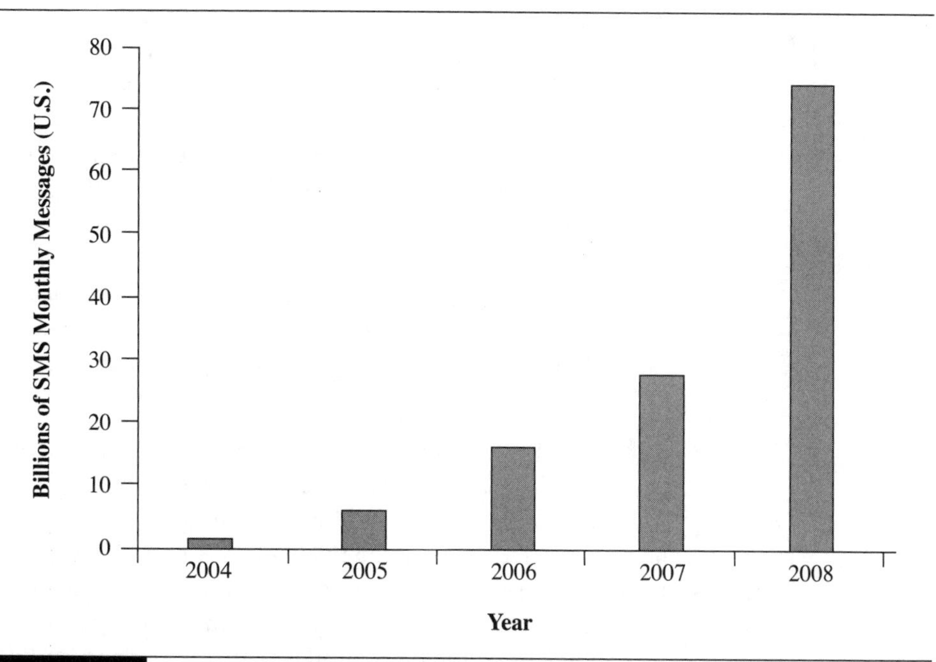

FIGURE 6-4 Explosive growth in texting

(www.gao.gov/new.items/d07149.pdf). The GAO makes two observations of interest to us here:

> *The various initiatives that FRA has begun in the past year and a half to better target its oversight—by addressing the main causes of train accidents and better focusing inspections on problem areas—hold promise for bringing down the train accident rate, reducing injuries, and saving lives. Some initiatives, such as reporting of close call incidents, encourage the railroad industry to address safety problems before they result in accidents.*
>
> *Also, FRA targets inspections at locations on railroads' systems where accidents have occurred, among other factors, rather than overseeing whether railroads systematically identify and address safety risks that could lead to accidents.*

What about the potential for accidents possibly associated with texting? The following excerpt came from a Reuters web posting (www.reuters.com/article/idUSN0152835520081002):

> *The train driver blamed for the worst U.S. train crash in 15 years was sending and receiving text messages seconds before his crowded commuter train skipped a red light and collided head-on with a freight train, federal investigators said on Wednesday.*
>
> *The Metrolink commuter train plowed into a Union Pacific freight locomotive on September 12 in Chatsworth, California, killing 25 people and injuring 135 in the worst train accident since 1993.*

Regulations prohibiting train personnel to engage in various activities are typical of the usual lag between the law and technology. What is needed is effective invention to reduce the threat to safety posed by this "new" technology. What do you suggest?

- The man-in-the-middle attack on public key cryptography is sometimes explained by analogy to a peculiar postal delivery system via the following scenario: Consider that Party A wants to mail a very private document to Party B. The mail service between the two parties is considered to have integrity in that the threat to privacy is at most *passive*, and that means that the mail service personnel might only read something that passed through their hands. This situation is similar to that of the postcard. It has no guaranteed privacy and the sender benefits from accepting this exposure to the passive threat by a reduced price. The parties want, however, to be certain that there can be no passive inspection of the very private document, so they mail the document in a steel box with a hasp that accommodates a physical lock, such as a padlock.

A problem arises in that the parties do not share any secrets a priori, such as the keys or the combination to the padlock. They settle on the following security protocol, sketched in Figure 6-5:

1. Party A puts his lock on the box and sends the padlocked box to Party B.
2. Party B receives the box with A's lock. Party B cannot open the box because she lacks A's key.

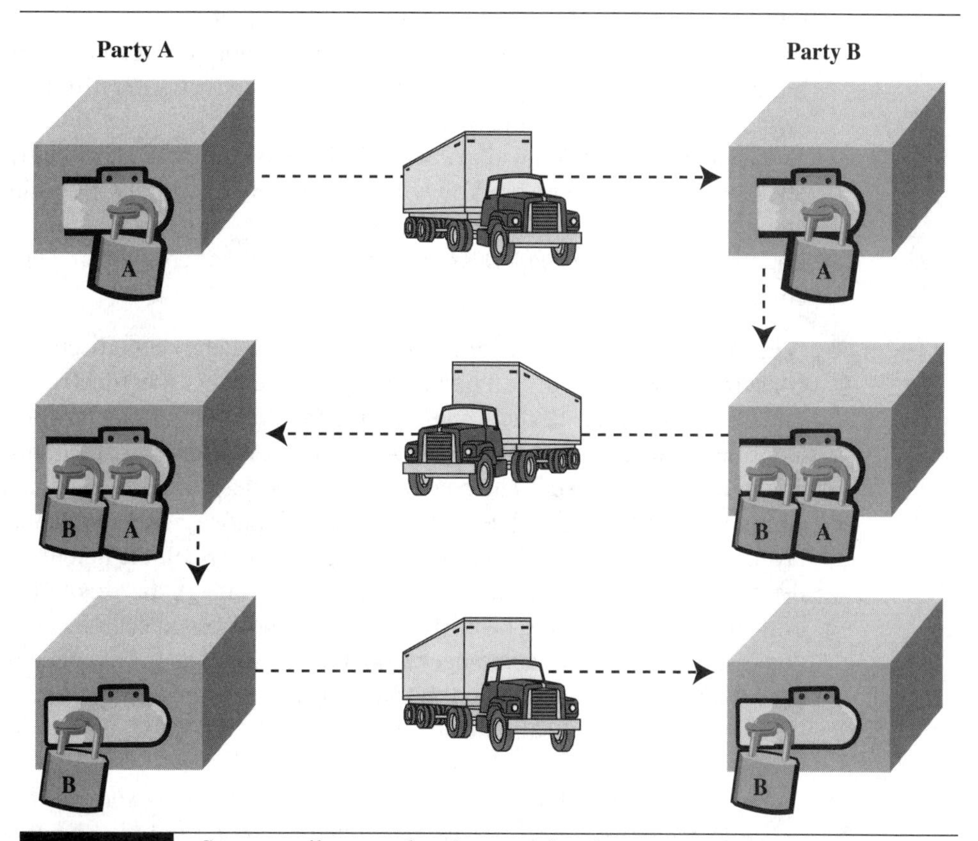

FIGURE 6-5 Secure mail protocol under a passive threat

3. Party B puts her lock on the box and sends the doubly padlocked box back to Party A.
4. Party A removes his lock and sends the singly padlocked box back to Party B.
5. Party B removes her own lock and removes and reads the private document.

Now suppose that the threat to integrity in the mail service moves along its dimension from *passive* to *active*, wherein an active threat can comprise interference with a transport, the insinuation of false items of mail, the denial or substitution of the mail, and many other such actions of that ilk. As depicted in Figure 6-6, by perturbing only one dimension of the system, we have an entirely different situation, with a maleficent agent, Party C, that can insert himself as a man in the middle and either make off with the secret, or, by symmetrically interposing himself, not only read the secret but have Party B believe that the secret message arrived safely from Party A and is shared between the two alone.

The man-in-the-middle attack will continue to require the best innovative efforts of inventors like you. You will find this attack virtually every time an attempt is made to improve transactional efficiency, and this includes such important activities as paperless purchasing, social networking, and cloud computing.

- A contemporary safety concern; There has been much recent discussion concerning the safety of cell phones and other devices, specifically the long-term effects of their non-ionizing radiation on human health. Consider patents such as U.S. 5787340 "Radiation shielding apparatus for communication device." The patent's abstract states: "A device for radio communication (PH), such as for example radiophone, includes a shielding layer (RFS) which under operating conditions is between the antenna and the user. The shielding layer (RFS) reduces electromagnetic irradiation of the user. The shielding layer (RFS) may be movable in such a manner that it serves a cover of some operational device such as for example the headphone, display and keyboard of the apparatus when it is not in use."

  The patent's specification reads in part:

  "The popularity of radiophones has been rapidly increasing during the last ten years. At the same time a belief of potential health hazards related to non-ionizing radiation has been increasing. The power

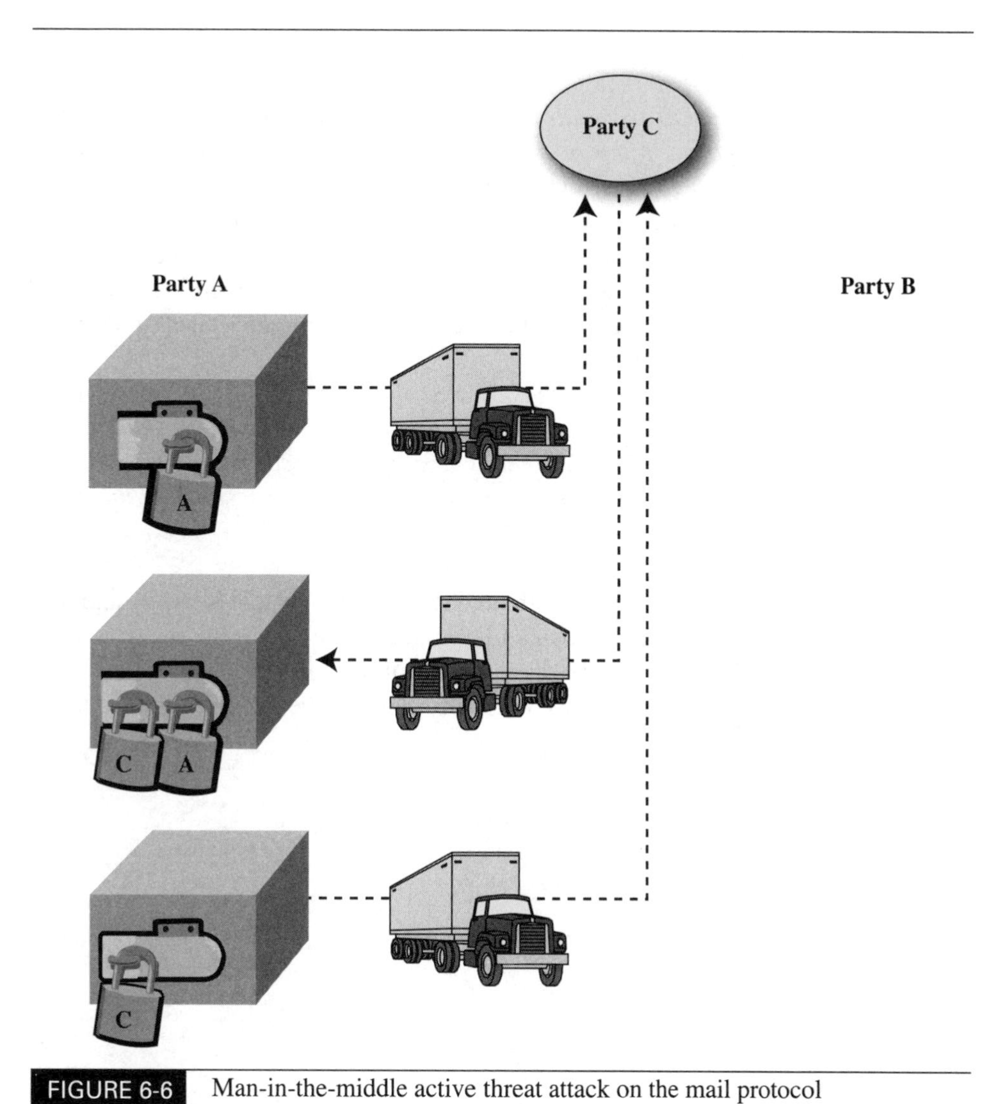

FIGURE 6-6 Man-in-the-middle active threat attack on the mail protocol

radiated by a radiophone is relatively low, typically few hundred milli-watts. On the other hand the antenna means of radiophones are few centimeters from the brain, the hearing organs and the organ of equilibrium. Although a direct heating effect could be left without further consideration it has been suggested that modulated radio frequency radiation induces changes in the electrical status i.e. in the ion balance of nerve cells. A continuous localized exposure to radio frequency irradiation has been suggested to weaken myelin sheets of cells and to eventually lead to an impairment of hearing capability, vertigo etc. It has been suggested that radio frequency irradiation may stimulate extra growth among supportive cells in the nerve system, which in the worst case it has been suggested could to a development of malignant tumor e.g. glioma from supportive cells. Although the consequences described above have not been scientifically verified, the uncertainty has some effects e.g. by reducing the speed of growth of the market of radiophones.

The invention avoids the drawbacks of the prior art and reduces the irradiation of the user, especially the brain and the nerve tissues. The invention and corresponding apparatus based thereon are characteized [sic] by what is set forth in the characterizing sections of the annexed claims."

Convene a meeting with an engineer and a patent attorney and discuss the pros and cons of such a filing under the two conditions: (i) it is determined that there is no harm resulting from the radiation and (ii) clear and convincing evidence is established that the radiation is harmful.

Chapter 7

Overcoming and Using Constraints

Hell, there are no rules here—we're trying to accomplish something.

—Thomas A. Edison

Quite often, invention will be required to work around a constraint, be it physical, cultural, geographical, or anything else that blocks your ability to achieve the required result. Constraints are often at the heart of inventive activity. An inventor loves constraints—they are so very often the fuel of Eureka! moments, and some of the greatest inventive achievements have been realized only after overcoming the most complicated and difficult constraints. Overcoming a constraint to create a meaningful solution can provide a great sense of accomplishment for an inventor. There is another side to the coin and that is the inventive opportunity to exploit a constraint. In this case the inventor works with the constraint instead of around it to arrive at a Eureka! moment.

In this chapter we will look at inventions that overcame constraints of energy and power limitations and of a difficult geography. We will also look at inventions that exploited constraints, one a mathematical constraint and the other a physical constraint, which happens to be gravity itself.

Energy Limitations in a Mechanical System

Many things can go a little bit wrong from time to time. And someone will always want to know what, where, and when these incidents occur. Unfortunately, life is sufficiently glutted with so many details that we cannot possibly tag and remember all of them. One example of such problems is the monitoring of mistreatment of transported goods, in particular, mechanical shocks such as impact shocks from vehicle mishaps and very rough roadways. These shocks can damage transported goods by outright breakage of components, loosening of interconnections, reduction of the lifetime of critical electronic components, and cosmetic damage to equipment housings. For insurance adjustment purposes and transportation quality assessments, it is desirable to monitor and categorize the occurrence of such impacts.

Of course, we are capable of building sensors and recorders, but powering such robotic detectives is a ubiquitous problem. Suppose, for example, that we need to provide power to a shock detector/recorder that's mounted on a railcar that's in service for many months or years. And suppose we can equip it with a battery of reasonable size that would last until it can be replaced. How do we deal with recording the nature and timing of mechanical shocks that occur over that period of time? To conserve power resources, our shock detector/recorder would need to wait until the rail car has hit something before it turns itself on to detect and record the incident, without missing the characteristics that define the beginning of the shock. In other words, by the time the shock event has been detected and turned the shock recorder on, the details of the shock, its duration and its intensity for example, would be past history and not recordable.

A great solution to this problem is taught in U.S. Patent 5585566, "Low-Power Shock Detector for Measuring Intermittent Shock Events." This invention provides an excellent example of an analogy aiding the Eureka! moment. The key to the invention is in the realization that a mechanical shock propagates rapidly but at a sonic rather than electrical speed through a mass. The mass is a mechanical delay line and "remembers" the shock as it passes through. As a kid, you may have tried the experiment of talking with a friend by using two cans attached via a string. The voice waveform travels along the string. The string is roughly analogous to the mass, as it "remembers" your voice until it gets to the receiving can.

Let's look at the patent, which is very straightforward and beautifully simple. The following is the motivation from the patent's background section:

It is desirable to provide an improved shock detector that allows for economical and reliable operation in a power-starved environment. It is also desirable to provide a shock detector that, notwithstanding its relatively low-power consumption over presently available shock detectors, is capable of measuring any shock event in a substantially accurate manner in order to provide to users meaningful information about the shock events.

The following is one embodiment of the patented inventive solutions:

The shock detector comprises a mechanical delay line, such as a torsional mechanical delay line or a spring and the like, to supply delayed output motion in response to input motion along a predetermined input motion axis. A transducer…is coupled to the mechanical delay line for converting the delayed output motion to an electrical signal representative of the delayed output motion. (Included is a sensing means to turn on the electronics.)

This invention has dealt with a limited *energy* source or budget. This particular constraint is repeatedly encountered. Motivated by our need to operate while being mobile, we carry more and more equipments about and they all require energy. We consistently need to consider how we will generate this energy or replace batteries or recharge them.

A close cousin to the above constraint is a power constraint. Power is energy per unit time. A power constraint is significantly different from an energy

> "Nearly every man who develops an idea works at it up to the point where it looks impossible, and then gets discouraged. That's not the place to become discouraged."
>
> —Thomas A. Edison

constraint but it is similarly universal in its appearances demanding its own set of Eureka! moments.

Power Limitations in the Railroad Industry

In the railroad industry, providing power for signals, rail continuity measurements, and other functions is a constant requirement. Generating power can be a tantalizing and frustrating problem. Batteries need care and replacement, solar cells can fail due to various environmental hazards such as bird droppings and vandalism, and generators need continuous fuel. A train that generates an enormous amount of power travels over the rails, yet none of that power is available to support roadside equipment. There is simply no straightforward way of transferring it.

As inventors, what can we do? Let's pull up to a grade-level crossing and think about it. A train rolls across the track in front of you. What do you see? You'll usually see one of the rails on a crosstie move up and down just a little as each car's wheel truck passes over it. Could this be the seed of a solution? Could this encourage a Eureka! moment? A very small amount of the train's kinetic energy is turned to heat by depressing that railroad tie. Suppose we could turn some of that energy into stored electricity?

In U.S. Patent 7148581, "Rail based electric power generation system," that is exactly what is taught. One embodiment shows a way to use the moving crosstie as an energy "pump." It uses a capacitor attached to the crosstie with plates that are separable and normally closed. When a railroad car's truck runs over the tie, the capacitor plates are separated, and after the truck has passed, a spring brings them together again into the closed position. The capacitor includes a very thin, high permittivity dielectric between the plates that can handle a high voltage. Before the truck separates the plates, the capacitor "pump" is "primed" with a small amount of charge that might be generated in a variety of ways, such as by a piezoelectric junction that is also compressed by the passage of the truck. After the capacitor has been primed with the small amount of charge, the capacitor is disconnected and the plates are opened by the truck passing over it. As the plates are separated, the capacitance drops dramatically from its value C_c (capacitance in closed position) to C_o (capacitance in open position). But because the charge remains constant, this drives the voltage up across the capacitor, increasing the electrical energy, which can be harvested by a rectifier and storage device.

The key to this inventive activity was to realize that there was a lot of kinetic energy present due to the moving train but in order to harvest any of it, it was necessary to search for a way to couple some of that energy to a stationary generator. The observation of the crosstie motion was key to finding and ultimately harvesting that coupling to the roadside equipment.

Energy Pumping Math Model

Mathematically, if *q* is the charge placed on the capacitor in the closed position, then the voltage across the capacitor in the closed position is $V_c = q/C_c$. When the capacitor plates are opened so that the capacitance goes to C_o, where $C_o << C_c$, the voltage across the capacitor in the open position is $V_o = V_c(C_c/C_o)$. The energy stored in the capacitor in the closed position is $E_c = \frac{1}{2}C_cV_c^2$ and the energy stored in the capacitor in the open position is $E_o = \frac{1}{2}C_oV_o^2$; thus the energy "pumping" action is $E_o = (C_c/C_o)E_c$.

Geographic Constraints

A unique type of constraint can be imposed by difficult geography, where severe conditions can motivate an unusual path to innovation. A great example of this is the development of the ALOHA protocol, one of the most important early computer networking innovations that supported the development of today's wireless networking.

Its story began in Hawaii, an archipelago comprising eight major islands and 750 miles of coastline. Scattered over the islands are the various campuses of the University of Hawaii. Back in the late 1960s and early 1970s, mainframe computing was the paradigm, with the mainframe computer located on the university's main campus. Back then, Hollerith cards, aka IBM cards or punched cards, were used to enter data into the mainframe. These cards were made of stiff paper and had 80 columns and 12 rows. Characters were entered by a series of punches in a column. The cards were ubiquitous familiar items and they were universally known among the computer literate. A student would typically compose a program by punching up a stack of cards and submitting them as a batch program to be run on the mainframe.

At that time, data was submitted in the framework of a remote job entry. Cards would be punched and sent to the student's program file that was being built from the punched cards. And therein lay the problem: How to get the card images into the mainframe computer system in a cost-efficient manner. The ocean and the distance that separated the computers on the various islands made this a particularly tough problem. In the book *Computer Communication Networks* (Prentice-Hall, 1973), UH Professor Norman Abramson addressed the problem and his ingenious, innovative solution:

> *The starting point in the design of THE ALOHA SYSTEM was the question, "Given the availability of a fixed amount of communications capacity, how does one employ this capacity to provide effective communication from remote users to a central machine?" Stated in these terms it became clear that the simple replacement of the wire communication channels of the common carriers by equivalent radio channels was not the answer.*
>
> *Remote terminals and Remote Job Entry (RJE) devices were introduced into the University of Hawaii system in 1969. By the end of 1971 approximately 50 terminals and RJE stations were connected using conventional dial-up and conventional leased-line communications. The cost of this communications system including telephone and leased-line charges, modems, and communication controllers had become a substantial portion of the University's computing budget.*

In other words, leased-line communications were terribly inefficient and therefore resoundingly cost-ineffective because they carried a single card image at a time, and the time lapse between card images was relatively long and random. Similarly, assigning a separate radio channel was also profligate in usage of time and bandwidth.

Abramson devised a system and protocol for communication that used only a small amount of bandwidth and allowed any user to transmit in the bandwidth whenever he wanted. The individual transmissions were short and randomly initiated. Most of them got through without interfering (colliding) with another transmission (packet). A feedback channel broadcast a successful packet receipt by the computer center. If a packet was lost because of noise or collision with another packet, no notice of successful receipt was broadcast and the broadcaster of a lost packet would retransmit the lost packet, but offset the start of its retransmission by a random time so that it would be less likely to collide again.

The ALOHA communications system, or *protocol* as it is better known, is ideal for handling this type of traffic, and it has become a staple in the communications networking art. It was a brilliant invention and development effort motivated by necessity and geographical hardship.

IBM Cards

IBM cards were at one time the lifeblood of a once-exploding technology. In the early 1960s, I chanced upon a creative photograph hung in one of the halls of a university. The picture consisted of a Bible resting on a flat surface and place marker that was a computer punched card. There was the ironic juxtaposition of an old object, the venerated Bible, and one of the most recent of the high tech inventions, the IBM punched card. And today? Well, few would fail to recognize the old Bible. But quite a few observers might ask about the "funny looking card." What we think of as cutting-edge technology today goes out of date so quickly!

Nonlinearity Constraint

Sometimes a constraint is an opportunity awaiting recognition and exploitation. Here we'll discuss a concept known as *reselling*, and you'll see how some folks made a lot of profit from a nonlinearity constraint.

One of the lessons in this example is that a nonlinear model or nonlinear behavioral curve can be a sign that the modeled system can be gamed in some fashion, or beware! It may game you.

The diagram shown in Figure 7-1 represents the land-line telephone structure of two towns. The local phone switches can connect any two subscribers within the same town. The problem becomes interesting when a subscriber in one town wants to be connected with a subscriber in the other town, however. The lines between the towns, the trunk lines, are expensive infrastructure, so the engineering challenge is to build only as many trunk lines as necessary to meet an acceptable figure of merit known as *quality of service* (QoS).

The following assumptions have worked well for telephone calls between human subscribers:

- Subscribers making calls do so independently of each other.
- The call generation rate is constant and identical over the subscribers.

Linear and Nonlinear

These are two important terms in engineering, economics, and many other sciences. They specify a type of relation between two variables. For example, a car's momentum is proportional to its speed. Momentum is a linear function of the speed. Kinetic energy, on the other hand, is proportional to the square of the speed and is therefore a nonlinear function of the speed.

The terms linear and nonlinear also characterize the graph plotting the relationship between the two variables. If the graph is a straight line, then the change in one variable is linear with the change in the other variable. If the graph shows a convex or concave curve, then the change in one variable is nonlinear with the change in the other variable.

- The duration of the calls is exponentially distributed.
- Calls are handled on a first-in-first-out (FIFO) basis.

The call generation rate is measured in Erlangs (after A. K. Erlang, who made seminal contributions to communication traffic modeling in the early 1900s). An Erlang is one call hour per hour. So a trunk that carries traffic for one hour carries an Erlang of traffic. (For example, if a single trunk carries 10 six-minute calls in one hour, it carries one Erlang. Note that to carry one Erlang in this fashion, the 10 six-minute calls would have to be precisely sequential.) The Erlang-B distribution models the case wherein blocked calls, calls originated while all the trunks are busy carrying other calls, are not queued and are dropped.

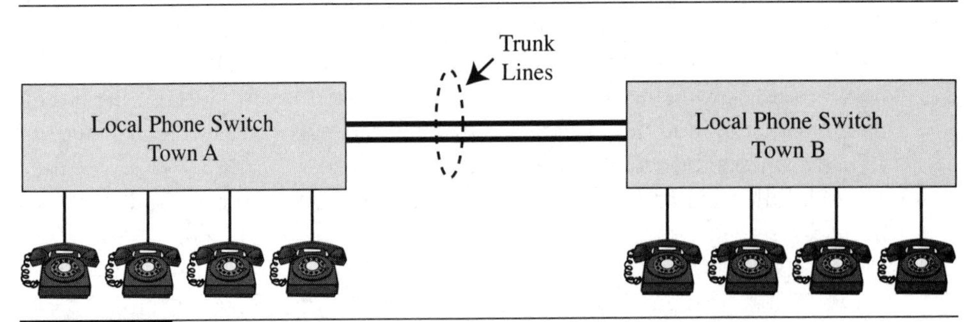

FIGURE 7-1 Local phone switches in two towns linked by trunk lines

NOTE

The probability of a blocked call with E-offered Erlangs and N trunks is

$$(E^N / N!) / \sum_{i=0}^{N} (E^i / i!)$$

.

Now here is the game. The model assumes we installed enough trunks to meet the required QoS: a call must not be blocked more than 1 percent of the time. The graph in Figure 7-2 shows the Erlang-B distribution for 1–10 offered Erlangs and 1–15 trunks. A line has been drawn at the 1 percent level of probability of blocking. The hint to the Eureka! moment is the nonlinear nature of the curves: Notice how they are not straight lines, but are curved. Remember that nonlinear behavior can signal an opportunity to game the system. So the question is, how do we turn this nonlinear behavior to our great advantage and profit?

Suppose we had E = 3 offered Erlangs to a telephone switch. From Figure 7-2, we see that we would need at least N = 8 trunks to have an acceptable QoS. Now suppose that another switch also handles 3 Erlangs of traffic and requires 8 trunks

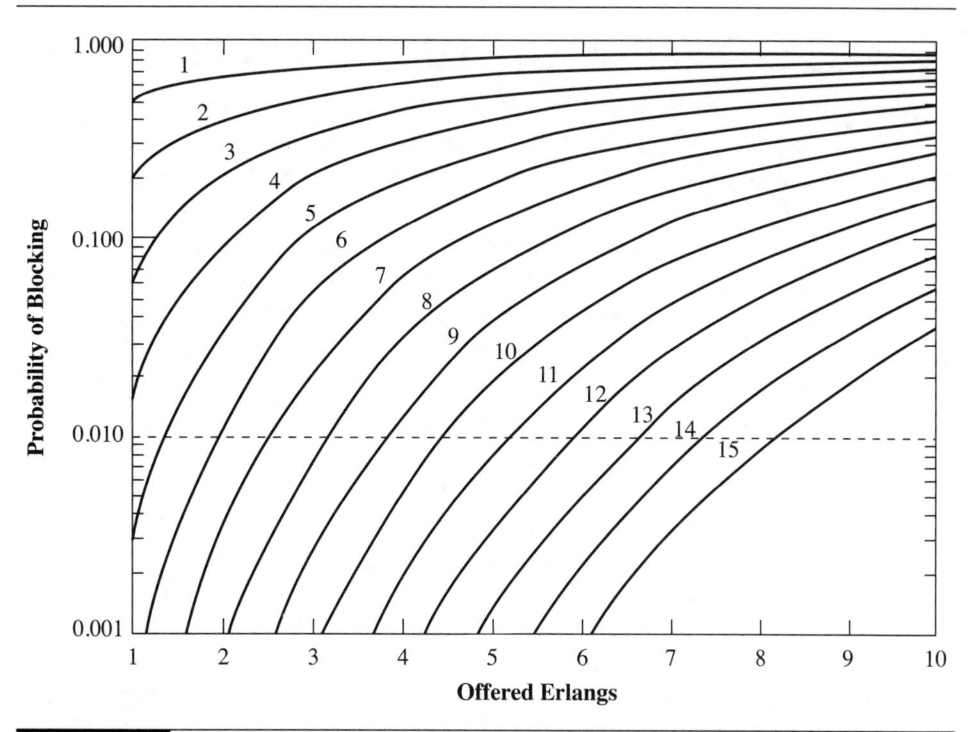

FIGURE 7-2 The probability of call blocking for 1–15 trunks and specified QoS

to meet the QoS. The curves in the graph are nonlinear. What would happen if we combined the traffic offered to both switches—that is, 6 Erlangs—and presented that to a single switch? How many trunk lines would be needed? Aha! We would need only 13 trunk lines to meet the QoS *and not* 16 trunk lines.

And this is what a reseller does. The reseller approaches all the subscribers and offers them the same QoS but at a lower rate, and the reseller saves by not having to pay for 3 of the 16 trunk lines. The reseller can therefore offer a less expensive service at the same QoS and also achieve a profit. This is a case in which understanding a little bit of business and applying someone else's mathematical results can yield a significant gain.

Reselling is just one example of taking advantage of a nonlinear relation. Many physical phenomena are highly nonlinear. For example, the lifetime for an incandescent light bulb is highly nonlinear if graphed against the voltage applied to the bulb. The reader is invited to explore the various issues and tradeoffs in using incandescent light bulbs designed for 130 volts on circuits supplying 120 volts.

Proving a Negative: Detecting a Null Condition

Our final example is quite unusual. Inventors are often motivated to create inventions that seek to detect the presence of something—a gas, a fire, a person, or an animal, for example. But sometimes we are asked to detect the *absence* of something, and this requirement can impose a significant challenge.

Reducing damage from impacts is a significant concern for all sorts of objects (including people)—consequently we have shoulder pads, knee pads, elbow pads, helmets, Styrofoam, and many other sorts of passive protectors. Living in our frenetic world exposes us to many opportunities for impact-based damage.

Active protectors, or intelligent protectors, are devices that deploy upon sensing an impact, such as vehicle airbags. It is interesting to observe the spread of active protectors to devices that are extensions of ourselves, such as our data storage and processing equipment that we carry about and fear to drop, especially treasures that have a spinning disk (that is, laptops). A disk crash caused by a fall can wipe out enormous amounts of data that can take significant resources of dollars and time to regenerate—if that is even possible. But how do we know when to park the read/write head before a fall results in a harsh and disk-fatal impact? This is an intriguing and important problem. Your first reaction as an inventor might be to try to harness and apply some of the technology associated with vehicle airbag deployment; it seems a reasonable place to start.

Let's look at an idea created at Apple Computer. It is taught in the company's U.S. Patent 6520013. Related sensors are first reviewed in the patent's background section, with the following included regarding the technology of airbag sensing:

> *A second kind of sensor is found in the unrelated automobile field. Sensors in this field are used to deploy various safety devices, such as airbags, whenever an accident occurs. Such sensors passively wait for an impact to occur and then rapidly deploy safety devices before a human's body impacts hard, bone-crushing surfaces within the automobile's interior cabin such as dashboards, windshields, and steering wheels. They cannot predict the possibility of an imminent impact, nor can they detect the absence of a gravitational field as some embodiments of the present invention can.*

So, what to do? How do we predict an impending impact by sensing the apparent loss of gravity that would be sensed in a free-fall? The answer is to look for something that changes if gravity is removed. And here it is! One embodiment of the invention taught in the Apple patent is illustrated in Figure 7-3, the patent's annotated figure 3.

The patent teaches using a small electrically conducting cantilever assembly consisting of a beam optionally terminated with an electrically conducting mass.

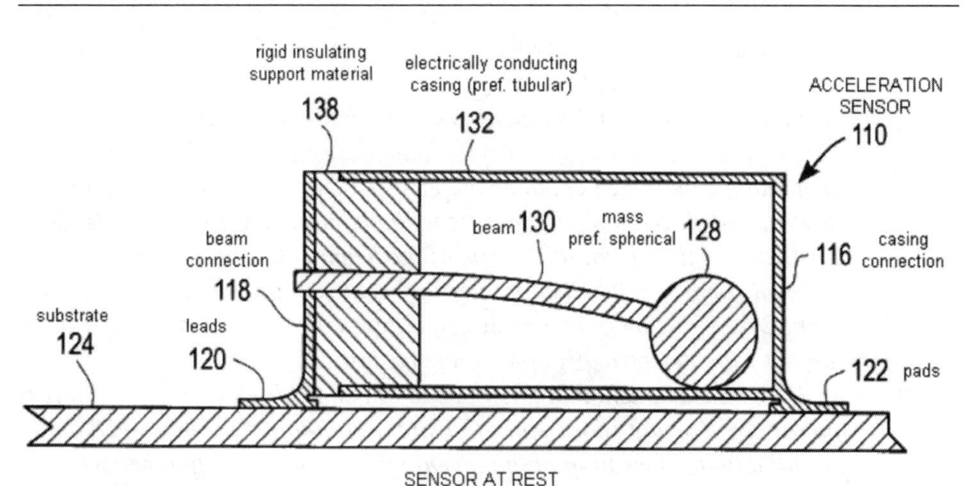

FIGURE 7-3 Figure 3 of U.S. Patent 6520013

Surrounding the cantilever structure is an electrically conductive housing. When gravity is present, the cantilever structure is deformed and bends so that it touches the housing. This completes an electrical circuit. Should gravity be removed, the cantilever structure would not be deformed to the side and the circuit would not be completed. Gravity is, of course, never removed but this is equivalent to the device entering free-fall. Follow the patent's specification with respect to an annotated copy of its figure 3:

> *Beam 130 may have any aspect ratio, meaning that beam 130 can have any cross-sectional shape. As exemplified in FIG. 3, beam 130 is flexible and electrically conductive. Preferably, the flexural constant of beam 130 is such that mass 128 contacts casing 132 when acted on by a gravitational force. Specifically, the flexural characteristics of the beam should be chosen so that two conditions are met:*
>
> 1. *The at rest gravitational force bends the beam, or beam/flexible mount combination, so that the beam or beam/mass makes electrical contact with the casing.*
> 2. *The lack of gravitational force during free fall allows the beam or flexible mount to straighten and break the electrical contact between the beam or beam/mass and the casing.*
>
> *In a preferred embodiment, insulating support material 138 is a rigid material such as glass, but other insulating materials such as plastic, epoxy, ceramic, etc. may also be used.*
>
> *In an exemplary embodiment, free end of the beam 130 may be weighted with a mass 128 to increase gravitational deflection and flex beam 130 such that the mass 128 contacts the electrically conductive casing 132. However, the invention can operate without mass 128. For example, in an illustrative embodiment, the shape of the beam 130, its dimensions, and the material comprising the beam 130 can be chosen such that the weight of the cantilevered portion of beam 130 itself flexes the free end of beam 130 into contact with a electrically conductive casing 132.*
>
> *If a mass is attached to the free end of beam 130, the mass 128 may take almost any size and shape since the size and shape of the mass 128 are not essential to the operation of the invention. It makes no difference whether the shape of the mass 128 is circular, squarish, polygonal, or triangular, as long as the mass is made of or carries an electrically conductive material and contacts electrically conductive casing 132 when the data storage device 103 is at rest.*

The preferable shape of the mass 128, as illustratively shown in the Figures is spherical.

According to one aspect of the present invention, the beam 130 and mass 128 are made of conductive materials or carry conductive means. Thus, electrical contact is made whenever either the free end of beam 130 or mass 128 touches casing 132. In this manner, the invention acts as an electrical switch, closed when at rest and open when in free fall. Beam 130 and mass 128 may be formed as one piece of electrically conductive material, or from separate pieces joined together by any suitable method, including, but not limited to, screwing, gluing, soldering, etc.

This last example hinges on exploiting the physical constraint of gravity. The inventors were required to detect the dangerous condition of free-fall and they did so by realizing that the condition of free-fall is equivalent to the absence of the gravity constraint and therefore the inventive path could be guided by detecting the absence of the gravity constraint.

Recap

Constraints form the envelope of the space within which we operate. If we wish to expand our operational space to accommodate new practices and new devices, we must often find a way to nullify, modify, or exploit a constraint. This challenge is readily accepted by the inventive spirit and the drive and determination of the inventor to accomplish something, to refuse to submit to the conventional rules of habit, is what promotes and procures some of the greatest inventions.

Discussions and Reflections

- Consider the sensor taught in U.S. Patent 6520013. What would it do on the space shuttle while it was in orbit? How would you propose to modify the sensor for work in orbit and on the ground?

- Steganography means simply "hidden writing." The constraint is not to be detected. Much of the successful practice of steganography is concerned with hiding secrets out in the open. Steganography has many forms. Some forms are chemical-based, such as the invisible writing that you might have learned about as a child: writing with lemon juice is made visible when exposed to the heat of a flame. Some forms use nothing but text.
  A famous case dates to the English Civil War in the 1600s. Sir John

Trevanion was imprisoned in Colcester Castle, facing imminent execution when the following note arrived for him (from http://math.ucdenver.edu/~wcherowi/courses/m5410/m5410cc.html):

Worthie Sir John:- Hope, that is ye beste comfort of ye afflicted, cannot much, I fear me, help you now. That I would saye to you, is this only: if ever I may be able to requite that I do owe you, stand not upon asking me. 'Tis not much that I can do: but what I can do, bee ye verie sure I wille. I knowe that, if dethe comes, if ordinary men fear it, it frights not you, accounting it for a high honor, to have such a rewarde of your loyalty. Pray yet that you may be spared this soe bitter, cup. I fear not that you will grudge any sufferings; only if bie submission you can turn them away, 'tis the part of a wise man. Tell me, an if you can, to do for you anythinge that you wolde have done. The general goes back on Wednesday. Restinge your servant to command. – R.T.

His jailers saw nothing amiss with the missive and passed it on to Sir John, who later asked to use the chapel for extended prayer and private contemplation. The jailers respected his wishes, but later on, when they entered the chapel to check on him, they found that he had fled. The key to his freedom was contained in the overtly innocent text and the message "Panel at east end of chapel slides" that emerges after sequentially taking the third letters after each punctuation mark.

In this example of steganography, a highly valuable message has been buried in a much larger message of little tactical or strategic value. The dimension of use for gaming the censor is simply the volume of packing text.

In electronics and modern communication systems, there are ample ways of passing extra bits of information in a manner that should evade casual censorship. Review the subliminal channel studied in Chapter 5 and spend some time devising some methods for using modern communication techniques to hide small messages. Then spend some time creating censorship methods to detect the methods you have just devised. This circle of thinking touches on some very important security problems today, and great demand exists for some outstanding Eureka! moments for their solution.

Chapter 8

Be Driven by the Bottom Line

Nobody is above the bottom line.

—Dr. Mark Grabb, Technology Leader for Analytics,
GE Global Research Center

If you are an inventor working for a company, it is essential that you align your inventive process with your company's bottom line or the customer's bottom line. Failure to channel your inventive process can result in many curious and cute but colossally unhelpful inventions. If you are overseeing or managing inventive talent, you must remember that inventors *have* to invent: it's part of their nature to create new things, and they won't stop for lack of direction. You must provide that direction in a meaningful and constructive way.

If you happen to be a lone inventor testing ideas in a home workshop, you should be sure that you understand exactly what you want out of your inventive efforts should they prove successful. It is not unusual to find a brilliant inventor who does not fully understand the business climate surrounding the invention and the need, means, and procedures required to protect his or her invention.

Rudderless Invention

The perils of "rudderless invention" became clear to me while I was a college undergraduate. A good friend came to visit; he was at a venerable business school as well befitted his natural talent. I regaled him with a story going around at that time about one of our brilliant chemistry students who had supposedly synthesized the attractant pheromone for the common housefly. According to the story, the student amused himself by mixing up a batch and applying it to statuary in the Boston Commons, attracting clouds of flies to the affected art. "What an idiot!" snapped my friend. "If he had put it on flypaper, he would have had a real moneymaker."

Recall Chapter 3's discussion of the gaming the system. Whatever way you find to game the system should always have bottom line improvement as its goal. Bottom-line oriented, or "guided," invention can also align to a significant degree with the bottom lines of your customers' companies. Some will say this point is painfully obvious—and perhaps it is. But do you keep it in focus as you go about your daily business?

The term "bottom line" is a catchall for identifying the goal for a company, client, or an invention. Quite often, the goal is simply to increase the profit of the invention's assignee or sponsor. This can be accomplished by inventing an entirely new product to sell or by improving an existing product to increase sales. Another important goal comprises the bottom lines of the sponsoring organization's customers. Again, this can be entirely profit motivated, or it can be motivated by a need for a device or method that has a particular fit within the customer's culture. This latter case is often a key requirement for nonprofit-oriented customers such as the government.

Whatever the bottom line, the inventor absolutely must discern and respect it. Failure to do this is likely to result in wasted time, money, and effort.

Let's consider some bottom line–driven applications of the inventive mindset.

Mature Technology Married to New Technology

Agribusiness has continually improved productivity, beginning with a remarkable jump in the 1950s. Continuing agricultural overproduction for national needs has created an economic model with price differentiation determining success over failure.

Techniques, methods, and the tools used to create even a slight increase in productivity can determine the survival or failure of an agricultural business. So, for example, if you wanted to sell agricultural machinery, your design engineers should know the theory and demands of precision or prescription farming. They should understand what is required for tilling, seeding, fertilizing, and harvesting. They should be able to design control systems to maintain very precise positioning and movements, even down to centimeter tolerances.

Consider the Global Positioning System (GPS) and its use in farm work. Caterpillar holds a significant number of patents involving the GPS. In October 2001, the company held 26 U.S. patents that contained the words "GPS" and "crop" in their specifications. Clearly, these are agricultural applications of GPS. But why are these large, relatively slow-moving machines good candidates for these small, high-tech devices? The answer lies in understanding some of Caterpillar's business—and, more succinctly, understanding agriculture.

In a January 1991 article in *GPS World*, Jeff Jacobsen, an extension specialist with Montana State University's Department of Plant and Soil Science, provided a clue: "GPS-based field navigation will be financially feasible in the future, possibly even in the next five to ten years." To understand the importance of that prediction, let's consider a little ag history.

Invention and Productivity in Agriculture

The march of agricultural productivity has passed through two basic periods. The first of these was the invention, institution, and refinement of labor-saving devices. Consider that in the days of hand tools, a farmer could reap about an acre of wheat per day with a scythe. By the turn of the twentieth century, much had been brought to bear on labor-multiplying devices compared to hand tools. Figure 8-1 shows the progression of labor savings in wheat farming for the years 1912–1984. As the required labor hours declined with increasing modernization, it became increasingly difficult to reduce them further.

The second wave of increased agricultural productivity was ushered in through increased crop yields, as shown in Figure 8-2, which shows wheat yield per acre. The net result was that agribusiness achieved continually higher productivity.

Combine higher productivity data and labor-saving data with continuing agricultural overproduction for national needs, and we realize an economic model that makes price differentiation a critical element in determining a farmer's success or failure. The techniques, methods, and apparatuses used to gain even a slight increase in productivity may determine a farming operation's survival.

One of the most important techniques that has helped improve agricultural operations and yields is the use of technology to help the farmer understand his agricultural land in extreme detail. All acres are not equal, and it is important that

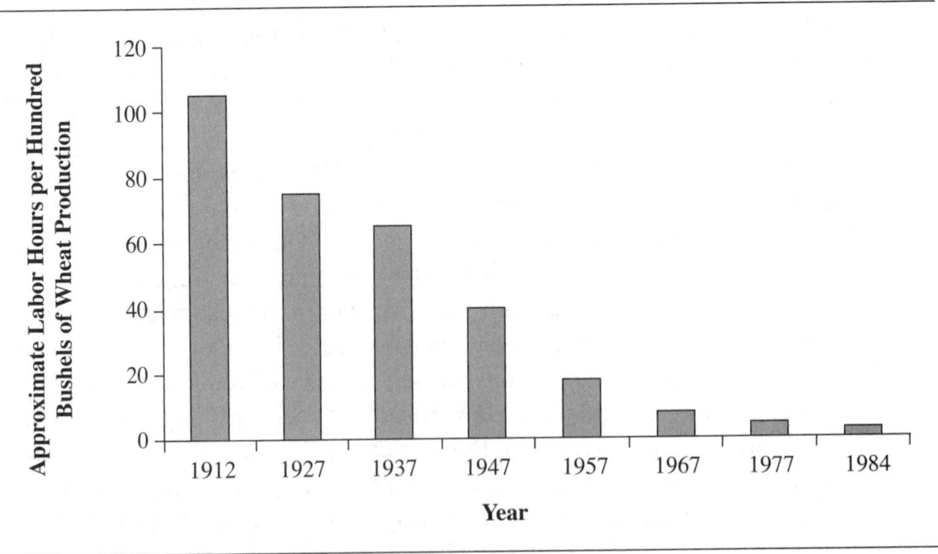

FIGURE 8-1 Labor savings in wheat farming

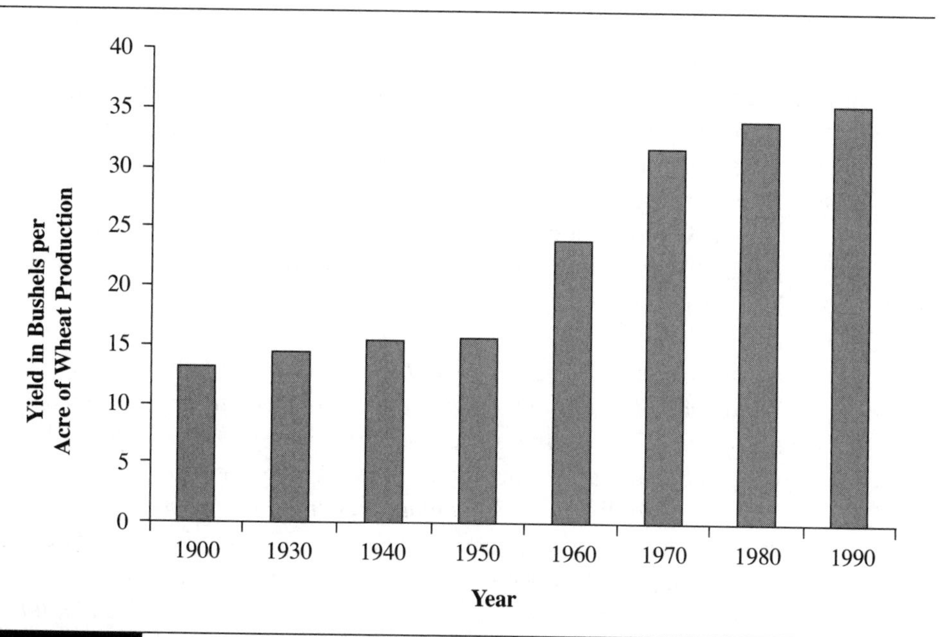

FIGURE 8-2 Yield per acre for wheat production

the farmer know his land and its characteristics. As the farmer works the land, he must also know the location where he is working, so that he can selectively apply fertilizer and seed, or other necessities as needed where it is needed. This is where GPS technology joins agriculture. A website maintained by Purdue's School of Agriculture explains this very well (http://www.agriculture.purdue.edu/ssmc/):

> *Site-specific management, also called precision farming, prescription farming, and even variable rate application technology, is an old idea that has been given new life by the advent of technologies based on global positioning systems (GPS). These GPS based technologies are used to tailor soil and crop management to match conditions at every location in a field.*

Caterpillar Patents

By now you should understand the correlation between GPS and Caterpillar farm machinery. Let's take a look at just how far the thinking had progressed as of the close of the millennium by considering a few Caterpillar patents.

Caterpillar was issued patents that clearly demonstrated the company's deep understanding of agricultural problems and its innovative application of GPS technology to them. As you know, the background section of a patent offers a comment on the prior art and attempts to set the stage for fulfilling the requirements for utility and novelty. Two of these patents are particularly noteworthy in this regard.

The first is U.S. Patent 6212862, filed in February 1999 and issued in April 2001, entitled "Method and Apparatus for Determining an Area of Harvested Crop." The background section provides a tutorial in and of itself:

> *Precision farming has evolved to address the needs of farmers to achieve optimal results in obtaining a high yield crop efficiently and economically. One aspect of precision farming is known as precision yield mapping. Yield mapping is essential to monitor the amount of yield, i.e., the yield per acre, in an agricultural field. Precision yield mapping enables a farmer to determine areas of the field which require special attention, such as applications of fertilizers and chemicals, which may not be needed in other portions of the field.*
>
> *Historically, yield per acre has been estimated by dividing the total yield by the number of acres traversed. This gives an average yield for the entire area, but does not determine yield in specific locations of interest. For example, a ten acre plot may yield a quantity of bushels of crop per acre, but the specific yield at any portion of the ten acre plot is not readily determined.*
>
> *With the advancements made in position determining technology, such as GPS position determining, yield per acre may be determined with respect to the position of a harvesting machine as the crop is harvested. This yield per acre determination may be mapped to a terrain database map to help establish a fairly accurate representation of areas in the field which require special attention due to low yields. However, these determinations are typically performed by multiplying the full width of the cutter bar on the harvesting machine by the distance the machine has traversed, thus determining an area of a rectangle covered by the cutter bar. The accuracy of the yield determinations are compromised in areas where the crop being harvested does not extend for the entire width of the cutter bar, such as around the perimeter of the field or when harvesting a narrow row of crop.*

U.S. Patent 6212862 defined a method and apparatus for yield mapping. Figure 8-3 shows a couple of the patent's figures. In the preferred embodiment, unit 210 is a GPS-based system.

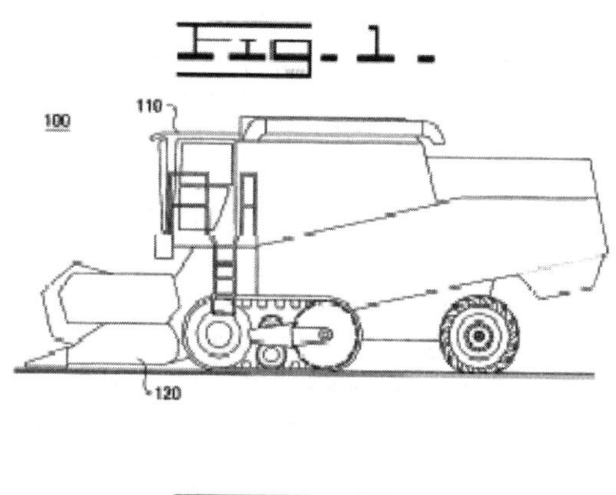

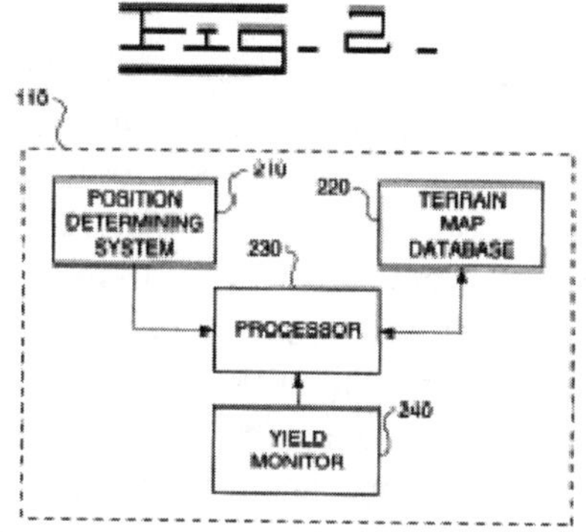

FIGURE 8-3 Figures of U.S. Patent 6212862

Now let's look at U.S. Patent 6236916, filed in March 1999 and issued in May 2001, entitled "Autoguidance System and Method for an Agricultural Machine." Again, the background section offers a great mini-tutorial:

> *Agricultural work often involves long, tedious hours in the fields, driving agricultural machines over all portions of each field to perform various tasks, such as seeding, fertilizing, harvesting, tilling, and the like. In addition, the paths to travel in the field must be carefully followed to cover the entire field efficiently and to avoid damage to the crop that is growing. Such work is highly fatiguing, and chance of error increases dramatically over time.*
>
> *The advent of modern technologies, such as GPS positioning systems, visual guidance systems, and machine control systems, have opened the door to automating many functions that formerly were required to be performed by the repetitious manual operations of individual workers.*

One such function involves the guidance control of mobile machines, such as agricultural machines for use in the fields. However, the tolerances required to guide an agricultural machine along rows of crops for extended periods of time create problems which make practical use of the above technologies difficult, if not impossible. For example, an agricultural machine in a field must traverse difficult terrain and yet maintain tolerances within centimeters to avoid damage to crops.

The addition of visual guidance systems, such as light bar guidance systems, offers some assistance in guiding an agricultural machine along these close tolerances. However, light bar guidance systems must be monitored closely, and maintaining this monitoring over long periods of time is difficult and tiring.

Figure 8-4 shows a couple of the patent's figures. Truly, this is an attempt to combine two very different technologies—one primarily mechanical and another electronic—to position bottom line growth for the agribusiness customer.

Agribusiness is just one example of a mature business that has benefited from the incorporation of new technology. In this respect, it is somewhat similar to the examples of large infrastructure improvements created through modern technology, but note an important difference. In our agribusiness examples, these innovations are not only improvements, but the improved agricultural machines are true combination inventions, and they are driven into existence by the customer's bottom line. In the case of Caterpillar, the company's innovations were developed with the farmer's bottom line clearly in sight. The farmer had reached the point of diminished returns from cutting labor costs and could now concentrate on his bottom-line enhancement through improved crop yields via precision farming.

Making Snow for the Ski Industry

A good example of invention applied to the customer's bottom line is also apparent in the ski industry.

Chapter 2 discussed improvement inventions, and there you learned that the ski industry is an "invention-persistent" business, because improvements continue year after year and at a significant rate. But beyond gear innovations, skiing absolutely requires one thing, and that, of course, is snow! Without snow, a ski area's bottom line suffers.

This concern led to the development of methods and machines that create artificial snow. Jeff Leich reviews the development of snowmaking in

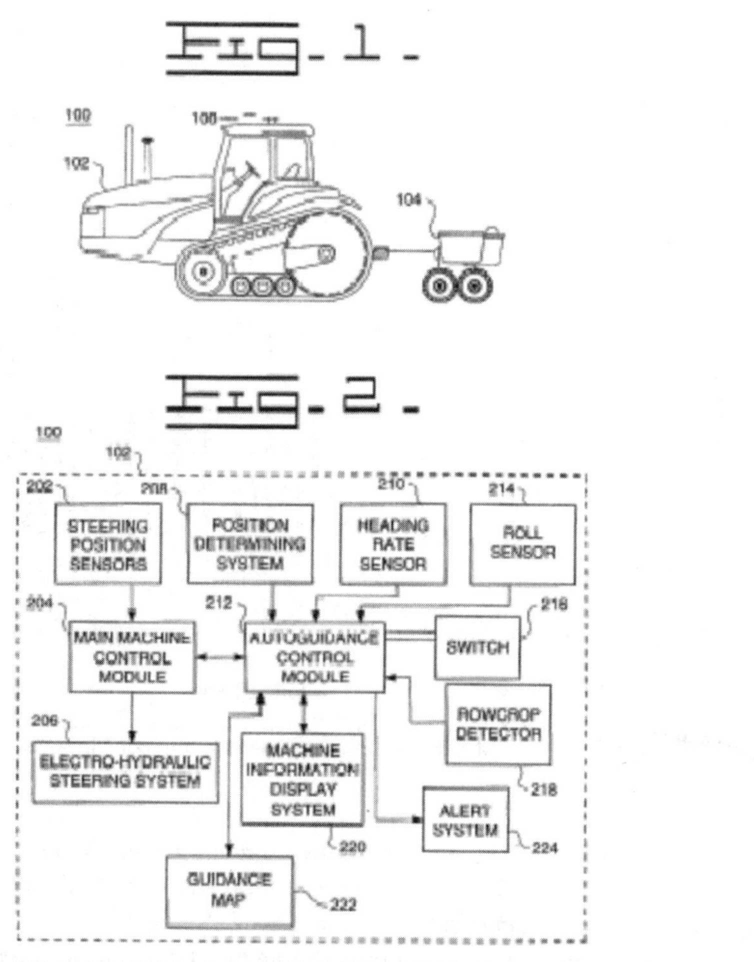

FIGURE 8-4 Figures of U.S. Patent 6236916

"Chronology of Snowmaking" (www.skimuseum.org/page.php?cid=doc141) and notes that the first major commercial snow-making installation was undertaken in 1952. Since then, the industry has been continually and aggressively improved as the graph in Figure 8-5 illustrates. It plots the number of artificial snow-making U.S. patents according to their filing year.

Snow-making had its start with Art Hunt, Wayne Pierce, and Dave Richey, who formed the Tey Corporation, the name deriving from the last letters of their three

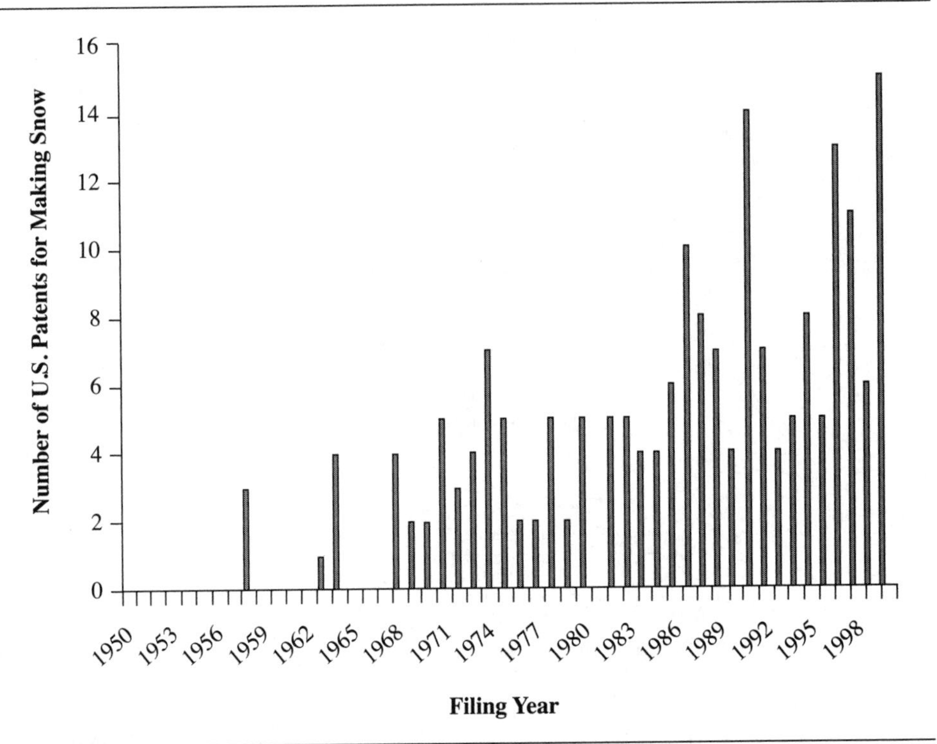

FIGURE 8-5 Number of U.S. patents for snow-making per filing year, 1950–98

last names. They proved their concept by producing a nozzle that mixed water and compressed air to form snow that was sprayed on the Mohawk Mountain in Cornwall, Connecticut, in December 1949. U.S. Patent 2676471 was filed by Pierce. The patent's introduction provides an excellent review of the situation and the bottom line problems that it overcame, thus motivating the invention's utility—being one of the basic requirements for patent issuance:

> *In recent years the enormous growth of interest in and the great increase in the number of people participating in winter sports has resulted in the growth or establishment of large-scale business to enterprises, the economic welfare of which is subject to the whims of the weather. On such winter sports is dependent a large industrial segment comprising ski area operators, ski lift operators, lodges, and manufacturers of winter sports equipment and the like. In recent years there have been marked variations in snowfall and accumulation in the north eastern portion of the United States and adjacent*

portions of Canada with the result, for example, that the ski seasons of 1948–1949 and 1949–1950 were extremely poor and occasioned considerable hardship upon those whose livelihood is derived from the winter sports business.

It has therefore become evident that could a means or method or apparatus be developed to economically produce snow upon slopes of ski areas, the financial hazards connected with those in the ski industry would be materially reduced.

Their perspective view illustrating their invention is shown in Figure 8-6. The valve-controlled outlets are numbered 17–23.

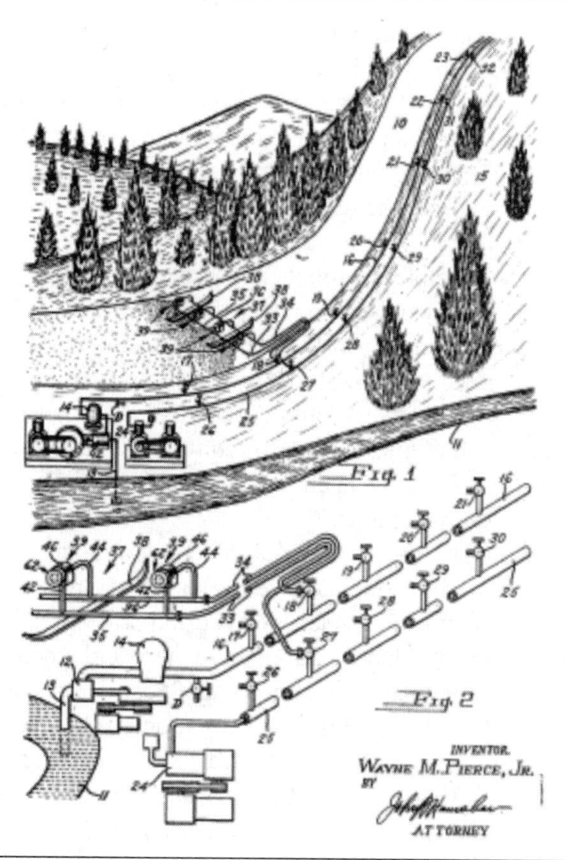

FIGURE 8-6 Drawing showing snow-making equipment protected by U.S. Patent 2676471

The invention of snow-making equipment illustrates that bottom-line improvement of an industry is a function of many different components. The continual growth of the skiing industry is dependent not just on skis and apparel but also on the length of the ski season, the number of days that the slopes may be used.

Making the Most of What You Already Have

New, large-scale efforts that are not merely improvements to an existing infrastructure, but entirely new functions, are a special challenge for the inventor. An inventor can often conceive of an innovation that will perform the desired functions, but the invention would require an enormous and costly effort. For large efforts especially, inventing "from scratch" is likely to make an inventor quite unpopular with managers and directors.

Instead of being the "tail that wags the dog," the inventor should try to envision approaches that make use of as much of the existing infrastructure as possible. Try to step back from what you envision as a problem's solution and think about what is already in place, owned, and maintained that might be used so that your invention meets not only the technical requirements of the bottom line but also its fiscal constraints.

Consider, for example, the impact of terrorism on the world and the inventive bottom lines created by these threats.

Terrorism has become a worldwide threat, as organizations and individuals become more innovative in their methods. In the United States, the bombings of the World Trade Center in 1993 and especially 2001 made it clear that terrorism was a formidable problem. Japan suffered its most serious terrorism attack in 1995 when the highly lethal nerve agent sarin was released in five coordinated attacks on Tokyo Metro subway trains.

Future poison gas attacks are a major source of concern for security experts and personnel. Security personnel need to know two main factors about an initiated or ongoing poison gas attack to curtail widespread disaster: its prompt detection and chemical identification, and where and when the gas is detected so that the spread of the volatile poisonous agent can be tracked and mapped.

Should inventors start building an infrastructure to detect, report, and map a spreading poison gas cloud? That would require a whole lot of sensors placed around every city, everywhere, and each sensor would need either a hardwired connection to a central command post or perhaps radios. If the latter is used, bandwidth would be required—and there's also the question of power. Do we hardwire the sensors and transmitters into the municipal power, or do we use batteries? If we use batteries, how do we schedule their maintenance or replacement? And the list of concerns goes on and on. The bottom line cost of building out and maintaining such an infrastructure is enormous. Inventing "from the bottom up" is probably not the way to go in this case.

But other options can make use of existing infrastructure. Recall the table showing POP scores for the technology linkage diagram in Chapter 5, the chapter on combination inventions. One of the biggest surprises is that the POP score between sensors and cell phones remained relatively high over the two time-separated searches. A colleague and I thought, and wrote, about this ("White Space Patenting," John E. Hershey and John F. Thompson, in Intellectual Property Today, Vol. 11, No. 8. August 2004). What could we do that would be useful, novel, and non-obvious with a sensor and a cell phone?

Eureka! All we need to do is to combine a biological/chemical sensor with a cell phone and add a little logic (hardware or software) so that detection of the presence of the contaminant would cause an automatic phone call to a 911-type message center for early warning. The message would identify the contaminant and the reporting phone's location. The operation would integrate naturally into cell phone usage. The cell phone user need not be trained in its use or even be aware that she is providing the mobile platform for detection and mapping. The process would take place totally automatically and would not depend upon any actions required by the cell phone user, or their health or lack thereof. Such a module could be added as standard issue and might eventually even be mandated—and that would be the path to regulation.

And invention is well on the march to get there. U.S. Patent application 20080088462, "Monitoring Using Cellular Phones," covers much of the required technology as shown in the following, taken from the objects and summary of the invention:

It is an object of the present invention to provide new and improved systems for obtaining information about people and their surrounding environment using cell phones, PDAs or similar electronic devices.

In order to achieve this object and possibly others, a method for obtaining information about a person in accordance with the invention includes providing the person with a portable device, arranging at least one sensor on the portable device, obtaining information about the person carrying the portable device or an environment around the person carrying the portable device using at least one sensor of the portable device without manual interaction, and transmitting the obtained information from the portable device to a remote facility. The portable device may be a cell phone or PDA.

The portable device may be provided with a system to obtain DGPS signals which are transmitted from the portable device to the remote facility to enable the location of the portable device to be determined from the DGPS signals at the remote facility. Alternatively, the location of the portable device may be provided by the portable device itself, if not determined from the information at the remote facility from information provided by the portable device, e.g., information relating to reception of signals by the portable device.

The information obtained by the sensor may be information about an environment around the portable device such as information about chemicals in the environment around the person carrying the portable device when the sensor is a chemical sensor.

Sensors arranged in association with the portable device may be one or more of a temperature sensor, radiation sensor, optical sensor, humidity sensor, chemical sensor, biochemical sensor, biological sensor, acceleration sensor, velocity sensors, displacement sensor, location sensor, vibration sensor, acoustic sensor and pressure sensor.

It is important for the inventor to understand the sponsor's bottom line. The preceding example emphasizes the additional importance for the inventor to survey and evaluate what is already in place that may be used. This is especially true for those projects that encompass great scale, i.e., large numbers of actors, large numbers of data sources, and large geographic areas.

Focusing on Bottom-Line Invention

How should we be looking for invention that integrates helpfully and naturally into the customer's world? I believe it is best to consider the customer and the customer's needs by answering questions in three sequential phases:

Phase 1

- What does the customer do?
- What does the customer want or need?

Phase 2

- What is the customer's culture?
- What are the customer's tools?

Phase 3

- What is most natural for the customer?
- What invention would be most easily integrated?

Let's try this approach focusing on a law enforcement customer, an officer who operates in a rapid and hazardous response such as raids. What bottom line should innovators heed here? Consider, for example, situations for which backup help is needed, or a brief message needs to be sent, or a team's position must be reported, especially during operations in remote or nearly impenetrable areas.

Based on just this information, and not heeding the three phases, what do we get if we turn the innovators loose? We get all sorts of little black boxes with dials, switches, meters, and cables. And what is the customer's requirement? Something that integrates easily and naturally into the agent's operational rhythm. Loading a gun and shooting is natural; setting switches and dials is foreign. So let's answer the questions of the three phrases and see what we get.

Phase 1

- **What does the customer do?** The customer is in law enforcement and conducts high-risk, high-energy operations.
- **What does the customer want/need?** The customer wants a way of summoning backup or briefly reporting a situation over a significant distance.

Phase 2

- **What is the customer's culture?** The customer is action oriented and non-technical.
- **What are the customer's tools?** Weapons and assault and arrest equipment.

Phase 3

- **What is most natural for the customer?** To carry and operate a minimum of offensive and defensive equipment.
- **What invention would be most easily integrated?** A specially designed bullet that can be inserted and operated (shot) as any other ammunition appropriate for the weapon.

This was the idea behind U.S. Patent 5381445, "Munitions Cartridge Transmitter." According to the patent abstract, this device is capable of emitting an electromagnetic signal after discharge from a cartridge-propelling device (a gun). It comprises a signal generator, an electromagnetic signal transmitter coupled to the generator, an antenna coupled to the transmitter, and a hollow cartridge for housing the generator, the transmitter, and the antenna. The transmitter is energized after discharge of the cartridge propelling device by a power source contained in the cartridge, as shown in Figure 8-7.

The line of sight in miles is proportional to the square root of the height of a transmitter above ground in feet. Thus, if the cartridge is shot straight up, its transmitted signal could be received and the bullet's location determined miles from the firing location, as sketched in Figure 8-8. The system integration is natural, and the customer's bottom line is respected.

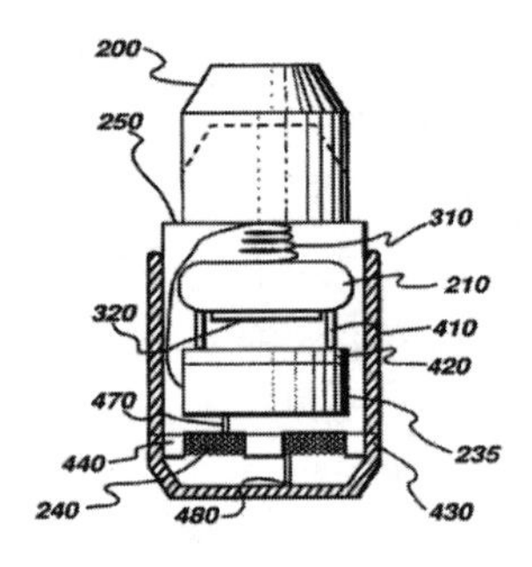

200 munitions cartridge transmitter

210 power supply

235 digital sequence generator & transmitter

240 antenna

250 hollow cartridge housing

310 electrically conductive retaining spring

320 positive electrode

410 insulating spacer

420 contact

430 sabot casing or jacket

440 bottom end of the cartridge

470 wire connected to bottom end of cartridge

480 mechanical attachment to casing

FIGURE 8-7 Munitions cartridge transmitter

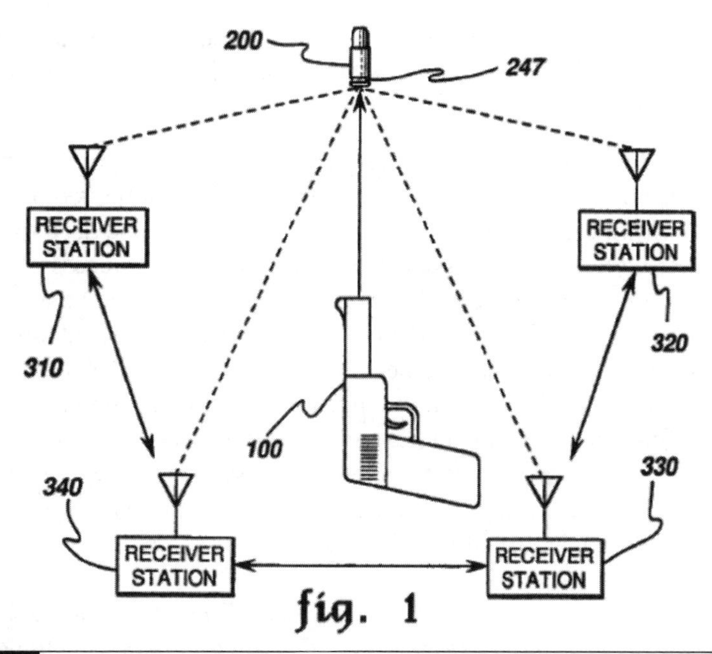

FIGURE 8-8 Operation of the munitions cartridge transmitter

Recap

A successful inventor is more than just an inventor who produces a new, useful, and nonobvious device, system, or method. A *successful* inventor knows his or her employer's or customer's bottom line. A *successful* inventor is guided by what is important to the user who will benefit from the invention. A *successful* inventor conducts research and experimentation while always respecting the sensitivities associated with the bottom line. These sensitivities can span the entire gamut of business; they can be cost concerns, ease of operational integration, or even societal and cultural concerns.

Discussions and Reflections

- Consider agricultural productivity beyond GPS. How might biotechnology impact crop productivity? Consider various scenarios, responses, and devise some "back-of-the-envelope" models. Is biotechnology a disruptive technology for agribusiness?

- List your important customers. What technologies are most important to their continued growth? What technologies are potentially most disruptive? What is the state of these technologies? Who is monitoring these technologies for you?
- IBM is concerned about respecting customer privacy. A contemporary technology that has become pervasive is that of the radio frequency identification (RFID) tag attached to an article and used to identify and track that article at a distance. Some tags allow successful interrogation at significant distances. Some customers are concerned that their purchases might be examined and identified by a party whose interests are not aligned with those of the customer and the retail chain. IBM addressed this concern with an approach that was ingenious in its effectiveness and its simplicity. Examine the following extracted portions from U.S. Patent 7253734, "System and method for altering or disabling RFID tags," and then attempt to answer the questions posed three phases discussed earlier in the chapter.

  From the patent's background:

  Large scale retailers and their suppliers are pursuing Radio Frequency Identification, RFID, tagging for supply chain tracking of goods. Demonstrations of RFID for item tagging will lead to point of sale check out and data collection. For the item tagging application, RFID tags are attached to some part of an item that is being inventoried or is for sale. The attachment may be such that the tag is not visible since the tag may be placed within a container section of the item or packaging material of the item. Removal of the tag after it is no longer useful can become difficult if not impossible for many practical situations. Thus, the tag will in many cases remain attached to the item that has been sold to a customer. This makes it possible for the tag to be read after the point of sale. This in turn leads to a question of the privacy of the purchaser or customer. The issue of privacy is of utmost concern.

  From the patent's summary:

  Another aspect of the invention provides a resonant tag kit comprising: a resonant tag; the resonant tag comprising an antenna; and an arrangement for reducing a read range of the resonant tag via physically compromising at least a portion of the antenna.

From the patent's detailed description, a figure (Figure 8-9)

> *shows a second embodiment **400**. Perforations **440** such as those used to enable the separation of postage stamps from each other are manufactured into the antenna and its substrate in such a manner that a separation along a line of perforations, separates the antenna **410** from the chip **420**, or a sufficient portion of the antenna from itself so as to disable the RFID tag **405**. A single or double line of perforations may be designed into the structure. A pull tab **450** may be added to facilitate the separation. In this embodiment a double line of perforations is employed. Thus, the consumer or a check-out attendant in a retail establishment, may perform the separation operation to disable the tag. The tag is open for xvisual observation for the confirmation of the disabling of the tag.*

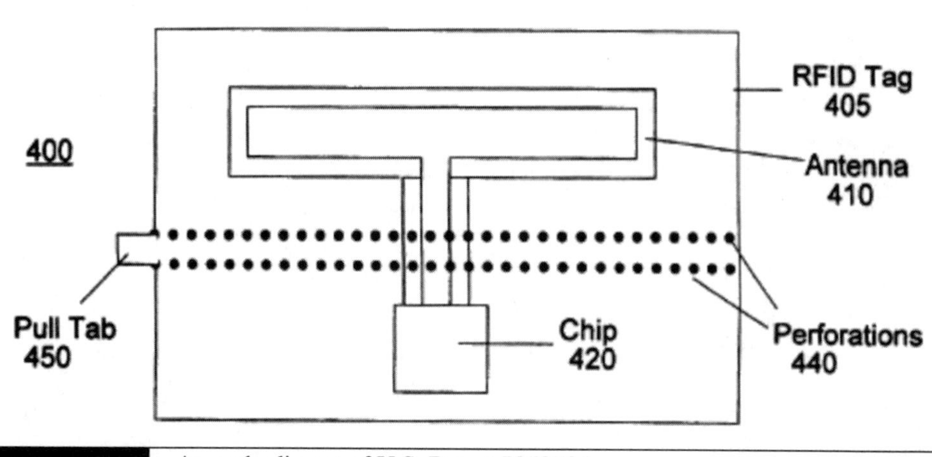

FIGURE 8-9 An embodiment of U.S. Patent 7253734

Part III

Appendixes

Appendix A

Patents: Mileposts of Invention

Let's take a closer look at patents, what they are and what they do for us. Understanding patents is important for two critical reasons

- *They are a deed to a type of intellectual property.* They convey rights that can be very valuable.
- *They are mileposts of innovation.* They are publicly maintained records of progress that in many cases have made life easier and more enjoyable, and they have strengthened the societal fabric in matters of productivity and safety, and preserved a competitive advantage.

There is much lore about patents and they are not generally well understood. The goal of this appendix is to help the reader with little technical and legal education understand and appreciate a patent and the patent process.

The Patent Art

The basis for a U.S. patent goes way back to the bedrock of the U.S. Constitution. In Article 1, Section 8, we read, "The Congress shall have Power to promote the Progress of Science and useful Arts, by securing for limited times to Authors and Inventors the exclusive Right to their respective Writings and Discoveries." Congress has chosen to do this by creating the U.S. patent system.

Now some folks think of a patent as a monopoly, one of the last monopolies left alone by the government. And because a patent can be extremely valuable, it is often referred to as "golden monopoly." But this terminology is not really correct. A patent, more correctly, conveys rights; yet a patent does not grant any positive rights to the holder. Somewhat surprisingly, perhaps, it does not ensure that the holder can practice the patent. Instead, the patent is a series of fences that prohibit all others from making, using, selling, importing, or offering for sale that which is patented.

To be granted a patent, an inventor must first pass a three-pronged test:

- The invention must be utile—that is, it must have some beneficial use. It must also be statutorily patentable material. Examples of excluded items are abstract ideas, algorithms, and nuclear weapons. But flamethrowers, bullets, smart bombs, and even genetically engineered microorganisms are fair game.
- The invention must not have been invented and known previously.
- The invention must not be obvious.

When an inventor or patent counsel speaks of a "101 rejection" of an invention, it is a reference to the first of these prongs from section §101 of title 35 of the United States Code. In conversation, patent folk often speak of these sections phonetically as "one-oh-one," "one-oh-two," and so on. As stated by the U.S. Patent and Trademark Office (USPTO), the government organ that examines patent applications and issues patents, "The subject matter of the invention or discovery must come within the boundaries set forth by 35U.S.C.101, which permits patents to be granted only for any new and useful process, machine, manufacture, or composition of matter, or any new and useful improvement thereof."

These rejections used to be relatively rare. One place they would crop up every so often was in respect to chemical patent applications. A chemist would invent a new molecule (known in the patent trade as a "new composition of matter") and be enamored with its structure or some such, file a patent application, and then receive a §101 rejection because there was no clear utility to the marvelous molecule. It might have been "really neat," but that was deemed insufficiently dispositive.

An Ad Hoc Rebuttal

A friend used to be chief patent counsel at a humungous electronics company. He had a method for dealing with §101 patent rejections in the chemical arts. He would advise the chemist who had invented the "really neat" molecule to whip up a batch or beakerful of it, go outside, and pour it on the lawn. If the patch of lawn died, its utility was herbicidal. If it grew, voila! It was a fertilizer.

The §102 rejection can be for a host of reasons, but they all spell a condition known in the trade as "prior art," and they mean that the patent examiner believes there is reasonably discoverable evidence that the invention is old news. This might be the case because of a patent that covers the same invention; a publication that discloses, or "teaches," the same; or the fact that the invention is publicly practiced. The wheel is a good example.

Finally, the §103 rejection deals with the admittedly fuzzy area of *non-obviousness*. What the examiner is saying here is that the inventor's invention would be obvious to someone who worked in the inventor's field. The one-oh-three rejection can involve mental gymnastics even for someone with a labyrinthine mind. The reason is that the examiner can argue nonobviousness in light of a single patent or publication, *or* the examiner can combine various patents and publications and claim that such a combination would have been obvious.

The one-oh-three rejection has undergone some serious scrutiny as of late and the Supreme Court has weighed in.

Satisfying the obviousness requirement has become an increasingly difficult hurdle. One-oh-three demands that an invention not be *obvious*, and obviousness extends to combinations of prior art—the sort of arguments that say, "Your invention is A plus B, but because A is out there and so is B, why isn't it obvious that one would naturally think of combining A and B to arrive at your invention?"

But it's getting even worse. Engineers practice and invent in what is known as the *predictive arts* as opposed to the chemists and bioscientists who often do not know exactly what will happen when they combine compounds or synthesize new products, and for that reason chemical and bioscientific art space is termed the *nonpredictive arts*. But how do engineers or anybody else determine what is obvious? After all, an engineer commissioned to build a bridge wouldn't say, "Well, here it is. I hope it'll hold up. I used a novel approach and it's different from any previous design." No, there would be simulations and demonstrable applications of known and proven engineering design principles. So where does a patentable idea in the predictive arts come from?

In 1941, the Supreme Court got into the act, and Justice Douglas wrote that a "flash of genius" was required to vet an invention for patent: "That is to say, the new device, however useful it may be, must reveal the flash of creative genius, not merely the skill of the calling. If it fails, it has not established its right to a private grant on the public domain." This doctrine was overturned in 1952 by a patent statute that declared, "Patentability shall not be negatived by the manner in which the invention was made."

Recently, the Supreme Court has again been revisiting the interpretation of patent law. In one of the latest salvos we find the pronouncement: "[The] combination of familiar elements according to known methods is likely to be obvious when it does no more than yield predictable results" (*KSR v Teleflex*, 550 U.S. 398, 2007).

At first sight, this is a bit chilling to inventors, but the court's opinion actually contains some surprising and even refreshing content. Although a judgment of obviousness is, in the end, legally determined, the court suggests that the definition, or rather the interpretation of obviousness, is specifically de-linked from acute and inflexible tests and is to be engaged in a more visceral confrontation, even to the point of invoking *common sense*. The court sums up the nub of the problem as follows:

> *Granting patent protection to advances that would occur in the ordinary course without real innovation retards progress and may, for patents*

combining previously known elements, deprive prior inventions of their value or utility.

The court also said the following:

It is common sense that familiar items may have obvious uses beyond their primary purposes, and a person of ordinary skill often will be able to fit the teachings of multiple patents together like pieces of a puzzle.

When there is a design need or market pressure to solve a problem and there are a finite number of identified, predictable solutions, a person of ordinary skill in the art has good reason to pursue the known options within his or her technical grasp. If this leads to the anticipated success, it is likely the product not of innovation but of ordinary skill and common sense.

The effects of the KSR case are being felt. A relatively recent case (*Leapfrog Enterprises, Inc. v Fisher-Price, Inc.*, 06-1402, Fed. Cir. 2007) decided by the U.S. Court of Appeals ruled that a patented invention did not deserve its patent because the inventive step was, in essence, to replace mechanical parts of a prior art device with electronic componentry. So, to qualify for a patent, an improvement invention needs to be a bit more than just going to an electronics store to buy the latest components to replace or upgrade existing modules or functions. The inventor will probably need to add functionality or do things in a different way.

But let's say that the inventor passes the three-pronged test. There is yet another hurdle to jump before the patent will be allowed. This hurdle has as its basis the consideration for the contract of the patent. The government insists that in exchange for the rights granted by the patent, for the protective fences to be put up, the inventor must teach the public how best to practice the invention so that when the patent protection expires (20 years from patent filing), the public will know how to make or use the fruits of inventor inspiration.

It is important, especially for technical folks such as scientists and engineers, to realize that a patent is not a technical paper. It is in fact something quite different. Scientists search for the connection between cause and effect, and their search is one of the noblest of undertakings. It's a tremendous achievement to tie together two events, perhaps widely separated in time or space, or both. To recognize that such an attempt should even be made is in and of itself a milestone. It is perhaps at the heart of ethical experience and behavior to assume it as a duty, for it is only in discerning the connecting threads of our life's fabric that we can make choices responsibly respecting others.

But a patent does not necessarily result from a deep understanding of principles. Instead, its value emerges from its practical teaching about how to build or do something of real value, even though the underlying principles of operation are not known, and might even be misunderstood. Remember learning of Genghis Khan, the famous Mongol ruler who lived from about 1162 to 1227? He was a highly successful military commander, and he promoted order among his people. In a set of laws he prescribed is one that commands, "Do not bathe or wash clothes in running water during thunder." What an absolutely marvelous example of dealing with a real danger without understanding anything about the underlying mechanisms. And there is an analogy to patents. Again, you don't have to understand everything about something that is valuable to teach, and perhaps patent.

Getting Started

Let's jump right in and look at expired U.S. Patent 6004596 "Sealed Crustless Sandwich," which is reproduced in Appendix B. This particular patent has attracted a lot of attention for various reasons, but I believe it is an excellent example for us because the subject matter is not complex, almost everyone will be familiar with the prior art, and, in my opinion, it is technically well constructed.

The patent's first page imparts a lot of information. It gives the title of the invention, in this case "Sealed Crustless Sandwich." It gives the patent number and the date it issued, Dec. 21, 1999. It reveals the assignee (the owner at the time of filing), Menusaver, Inc. (The patent rights might be subsequently sold or licensed. In this case, the Smucker Company acquired the rights and for a time successfully protected a product by these rights, up through a successful infringement trial.) The filing date for the patent is also shown, as are the inventors.

NOTE *Inventorship is a critical item to the USPTO. If inventorship is not correctly recorded—if someone who was an inventor is not listed or if someone is listed as an inventor who was not an inventor—the patent's validity can be assailed. This might be why a well-polished patent attorney will sometimes refer to listed inventors not as inventors but rather as named inventors. Inventorship is discussed in the Appendix C.*

The first page also lists references, patents, and other publications ("50 Great Sandwiches") that the applicant wants to cite or that are cited by the examiner. These references are relevant to *prior art*. Remember the 102-test—novelty? By citing a reference, the patent examiner is saying that the USPTO believes that the material in the patent is novel in light of the cited prior art.

The first page also carries the patent application number and the patent abstract, which is required. It cannot exceed 150 words, and its purpose is to enable the USPTO and the public generally to determine quickly from a cursory inspection the nature and gist of the technical disclosure. And, finally, the first page generally has a drawing or figure. The figure is selected by the examiner, and unless there is a caption to the effect that the figure is prior art, the figure is to represent or relate to an embodiment of the invention.

The next few pages contain the patent figures. Note that the figure chosen for the first page of our example patent is the patent's figure 3.

The remainder of the patent is generally formatted as text of two columns per page with the column number posted in bold at the column head. There is no hard and fast rule for patent format or sections beyond the requirement that there be a specification, an abstract, a set of claims at the end of the specification and technically a part of it, and figures, if the patent admits illustration.

The sealed crustless sandwich patent starts out with a background section. I have always enjoyed reading background sections, because they can review the prior art and can be great mini-tutorials on the general field of art. The inventor (more likely the patent preparer) must take care to keep all aspects of the invention out of the background section, because there is an implicit admission that any detail published therein is considered to be prior art, and any invention placed inadvertently therein could be, and has been, irrevocably surrendered as such.

So, if the background section is a bit of a minefield, why do patent preparers often include one? The answer is that it might motivate the utility of the invention. Remember the one-oh-one test? Why has this patent application been submitted? Why is this invention useful? Is there a need for this invention? What are the vexing unsolved problems in the art?

The sealed crustless sandwich patent's background does a great job at motivating utility. See lines 11–21 of column 1:

> *Many individuals enjoy sandwiches with meat or jelly like fillings between two conventional slices of bread. However, some individuals do not enjoy the outer crust associated with the conventional slices of bread and therefore take the time to tear away the outer crust from the desired soft inner portions of the bread. This outer crust portion is then thrown away and wasted. There is currently no method or device for baking bread without having an outer crust. Hence, there is a need for a convenient sandwich which does not have an outer crust and which is not prone to waste of the edible outer crust portions.*

One useful aspect of this invention is that of a crustless sandwich, responding to the need of many individuals. Now look at lines 26–31 as they recount some of the prior art:

There are numerous sandwich devices. For example, U.S. Pat. No. 3,690,898 to Partyka; U.S. Design Pat. No. 252,536 to Goglanian; U.S. Design Pat. No. 293,040 to Gagliardi; U.S. Design Pat. No. 317,672 to Presl; U.S. Design Pat. No. 318,360 to Sam; and U.S. Pat. No. 5,500,234 to Russo all of which are illustrative of such prior art.

If the prior art left no room for invention, there wouldn't be any utility in developing a new sandwich. So where does the prior art fail to cover a need? This brings us to lines 32–38:

While these sandwiches may be suitable for the particular purpose to which they address, they are not as suitable for providing a convenient sandwich without an outer crust which can be stored for long periods of time without a central filling from leaking outwardly.

The prior art does not teach a sandwich without an outer crust that retains an inner filling for extended periods of time, and this tells us of a further problem in the prior art—inner filling leakage. The picture in Figure A-1 illustrates the problem. A peanut butter and jelly sandwich was prepared with grape jelly on one side and peanut butter on the other. The sandwich was wrapped in plastic and sat for a couple of days. Notice the discoloration produced by the jelly's leaking through the bread. (Unappetizing to say the least!)

How does the new sandwich work to overcome the prior art problems? The detailed description of the invention must be written in such a way that someone "of ordinary skill in the art" can practice the invention—in this case, build a sandwich that will overcome the problems in the prior art.

Remember the few words about enablement? It is required of the patent applicant that the description enables a qualified practitioner to practice the invention. This is part of the contract between the government and the inventor. In effect, the government gives the inventor rights in exchange for the inventor's teaching not only a way to make the invention (the sandwich) but also the best way that the inventor knows at the time of filing. This best way is termed the *preferred embodiment*.

The onus for clear disclosure and teaching is spelled out in yet another section of patent law recited in 35USC112, spoken of affectionately as "one-twelve."

FIGURE A-1 Sandwich inner filling leakage problem

§112 has six paragraphs of requirements. They all are important, but the one that concerns us here, now, is paragraph 1:

> *The specification shall contain a written description of the invention, and of the manner and process of making and using it, in such full, clear, concise, and exact terms as to enable any person skilled in the art to which it pertains, or with which it is most nearly connected, to make and use the same, and shall set forth the best mode contemplated by the inventor of carrying out his invention.*

And, yes, if the patent examiner determines that the applicant has failed to meet this requirement, the examiner will issue a "one-twelve" rejection.

The applicant is free to add in any number of ways to practice the invention (to make the sandwich), and it is generally a good idea to include a number of methods, because it can help broaden the patent and increase its resistance to a "design-around effort." Remember that the best mode must be included in the different methods that are taught. (By the way, it is not necessary to identify specifically the best mode if more than one mode is taught.)

Looking at the '596 sandwich patent (when a patent number must be used more than once in a discussion, it is common to refer to the patent by its last three digits as long as there is no ambiguity), we see that the applicants called out the preferred embodiment in column 2. Let's look at just a portion of it, column 3, lines 8–22, along with figures 3 and 4, and comment as we go along.

As best shown in FIG. 4, the upper filling 30b is juxtaposed to a lower surface of the upper bread portion 22. The lower filling 30a is juxtaposed to an upper surface of the lower bread portion 20. Preferably, the upper filling 30b and the lower filling 30a do not extend into the crimped edge 26 since any foreign substance within the crimped edge 26 weakens the seal between the lower and upper bread portions 20, 22. The center filling 32 is positioned and sealed between the upper filling 30b and the lower filling 30a as shown in FIG. 4 of the drawings. The crimped edge 26 preferably has a plurality of depressions 28 formed into from the pressure points caused by the notched end 44 of the sleeve 42. The depressions 28 prevent the crimped edge 26 from separating thereby retaining the fillings 30a-b, 32 within.

Referring to Figure A-2, notice how precisely the language, in combination with the numbers and the pictures, describes the invention.

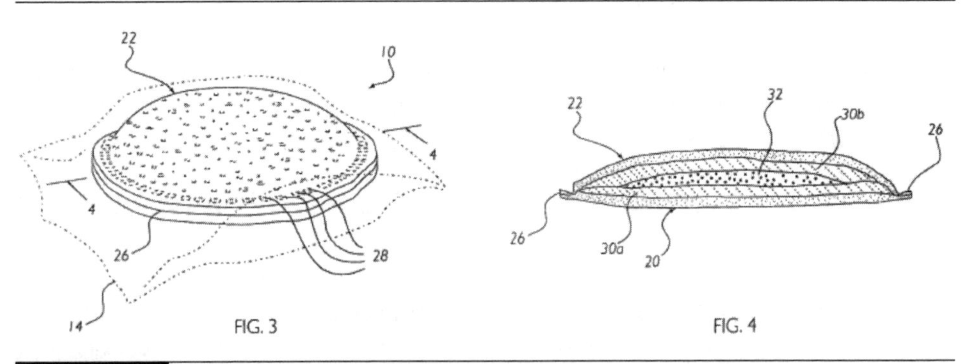

FIGURE A-2 Figures 3 and 4 of U.S. Patent 6004596

The Claims

At the end of the specification are the claims, perhaps the single most important part of the application, because the claims protect the invention: If it is covered in the claims, it is protected; if not, even though mentioned, or laboriously taught, in the specification leading up to the claims, it is material that is dedicated to the public and is not protected by the patent. Drafting the claims is the most difficult part of preparing a patent application. It takes great skill and tremendous verbal dexterity. Some say that a patent application is the most difficult legal instrument to draft.

Remember our dalliance with one-twelve paragraph 1? Remember that we said one-twelve had six paragraphs? Well, here is paragraph two: "The specification shall conclude with one or more claims particularly pointing out and distinctly claiming the subject matter which the applicant regards as his invention."

Let's take a look at the claims of the sandwich patent and see if we see the teachings of the specification within them. Claim 1 reads as follows:

> *A sealed crustless sandwich, comprising:*
>
> *a first bread layer having a first perimeter surface coplanar to a contact surface;*
>
> *at least one filling of an edible food juxtaposed to said contact surface;*
>
> *a second bread layer juxtaposed to said at least one filling opposite of said first bread layer, wherein said second bread layer includes a second perimeter surface similar to said first perimeter surface;*
>
> *a crimped edge directly between said first perimeter surface and said second perimeter surface for sealing said at least one filling between said first bread layer and said second bread layer;*
>
> *wherein a crust portion of said first bread layer and said second bread layer has been removed.*

The first part of the claim, "A sealed crustless sandwich, comprising:" is known as the preamble. The preamble is usually of very limited import. It usually serves to define the context within which the invention falls or is expected to be practiced. When you're reading a patent, don't assume that the preamble restricts or limits the invention. For example, if the preamble stated, "An engine for an aircraft comprising:," this might not rule out the patent protection for the same engine used in a power plant. For the preamble to restrict the claim's scope, the preamble must give meaning to the claim; it must, as one court opinion so eloquently stated, "breathe life" into the claim.

After the preamble, we encounter the elements of the claim. The invention protected by the patent must include all the elements, and, conversely, for you to infringe the patent, you must practice using all of the elements. This is commonly known as the *all elements rule*.

Claim 1 has four elements, and what is claimed is, essentially, two pieces of bread with one or more fillings, one of which is put on one of the pieces of bread. The edges of the two pieces of bread are sealed by a crimped edge, and, finally, some or all of the crust of the two pieces of bread is removed.

Claim 1 is called an *independent claim,* because it does not depend *on* or *from* another claim, unlike the second claim: "The sealed crustless sandwich of claim 1, wherein said crimped edge includes a plurality of spaced apart depressions for increasing a bond of said crimped edge."

A sandwich protected by this claim must have all the elements of claim 1 and must also have a crimped edge made by a plurality of depressions. (*Plurality* is a favorite word of patent practitioners and means *more than one*.)

Claim 2 is a dependent claim, depending from claim 1. It is important to grasp the distinction between these two types of claims. An independent claim often starts a chain of claims. In this patent there are ten claims, and two are independent. There are thus two chains of claims. If you take a look, you will see that claim 2 depends from claim 1, claim 3 depends from claim 2, and so on, to claim 8 that depends on claim 7. When a claim depends from another claim, the interpretation is that the latter claim has all the elements of the claim from which it depends, plus the elements of the latter claim. A series of dependent claims, such as we have here, is almost always a successive narrowing of the invention.

You're sometimes advised to think of an independent claim, say claim 1, as a bounded physical piece of property, or a white space of invention, with the boundary setting the limits of the claim. A claim dependent on claim 1 must have its entire boundary within the claim from which it depends, and its total area must be smaller than the claim from which it depends. If it were not smaller, it would cover the same ground or white space as the claim from which it depends and would not be differentiable from that claim. (Claim differentiation is an important doctrine in patent crafting.) The situation for the '596 patent can be envisioned with the sketch in Figure A-3.

And here's claim 3: "The sealed crustless sandwich of claim 2, wherein said crimped edge is a finite distance from said at least one filling for increasing said bond." What does this dependent claim add? Simply that it covers a sandwich wherein the crimping takes place further out on the sandwich than at least one of the sandwich fillings extends.

Next is claim 4:

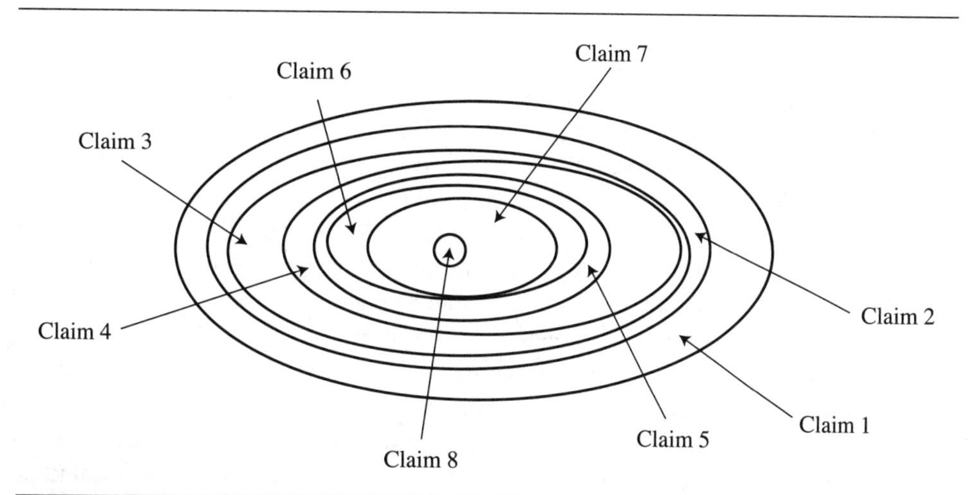

FIGURE A-3 Claim chain boundaries

The sealed crustless sandwich of claim 3, wherein said at least one filling comprises:

a first filling;

a second filling;

a third filling; and

wherein said second filling is completely surrounded by said first filling and said third filling for preventing said second filling from engaging said first bread layer and said second bread layer.

Now we're into multiple fillings with the restriction that the middle or second filling is kept from touching either of the two pieces of bread.

Claim 5: "The sealed crustless sandwich of claim 4, wherein said first filling and third filling have sealing characteristics." This is interesting. It seems we now also want to claim the possibility that the first and third fillings seal in the middle filling.

Claim 6:

The sealed crustless sandwich of claim 5, wherein:

said first filling is juxtaposed to said first bread layer;

said third filling is juxtaposed to said second bread layer; and

an outer edge of said first filling and said third filling are engaged to one another to form a reservoir for retaining said second filling in between.

The plot thickens. Now we want the two sealing fillings to encase the middle or second filling.

Claim 7: "The sealed crustless sandwich of claim 6, wherein said first filling and said third filling are comprised of peanut butter; and said second filling is comprised of a jelly." Aha! Look at Figure A-2. The two bread layers are numbered 20 and 22, the jelly is indicated by 32, and the sealing wrap-around peanut butter layers are indicated by 30a and 30b. Finally, 26 points out the crimped edge. Notice that if someone were to make a sandwich that was described and protected by claim 6, but the sandwich did not include peanut butter and jelly, there would be infringement of claims 1 through 6, but not infringement of claims 7 and 8.

Claim 8: "The sealed crustless sandwich of claim 7, wherein said crimped edge is formed into a substantially circular shape." Well, I guess we hadn't seen much about the sandwich shape until now. It could have been *traditionally* square. Under this dependent claim we are specifically calling out and protecting a round peanut butter and jelly sandwich.

Figure A-4 shows the cross-section of an actual sandwich (one of Smucker's own) prepared according to the patent's figure 4 and its claims.

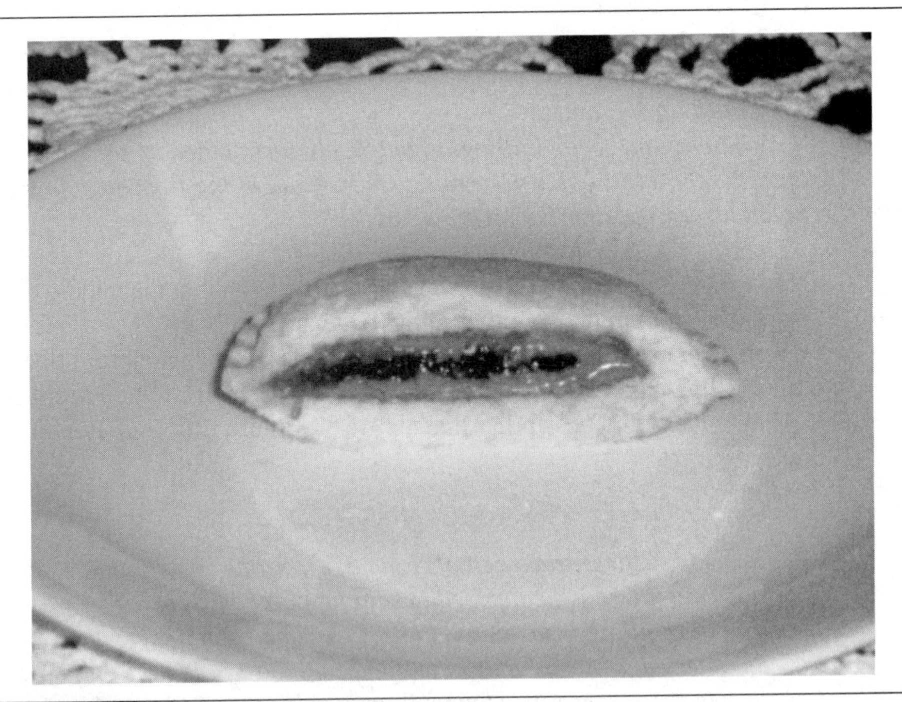

FIGURE A-4 Cross-section view of the patented sandwich

Did you happen to notice that each claim consists of precisely a single sentence? This is part of the required form for constructing a patent claim. This single sentence rule can admit some rather cumbersome reading. I used to know a patent attorney who was famous for using more than 1000 words in some of his claims.

Patentability, Infringement, and Design Arounds

We have spoken of what it takes for an invention to be patentable, but another analysis is possible: the analysis for infringement—that is, does your proposed invention infringe another's patent. And finally, if you do find yourself potentially infringing, how do you "design around" the other patent so that you have a patentable invention and do not infringe another's patent?

Consider, for example, that someone has patented a chair, and the claim for the chair comprises three elements appropriately linked together: a seat, legs, and a back. Now suppose that someone adds a set of rockers to the chair and thereby invents the rocking chair. The rocking chair patent claim has four elements: the seat, legs, back, and rockers.

Clearly, the rocking chair infringes, or "reads on," the prior art patent of the chair because it duplicates the chair's claim with respect to all of the chair's elements. But the rocking chair has the additional element of rockers that are not present in the chair's claim. The chair and the rocking chair are distinct inventions, and both are patentable, but as a result we have the interesting situation: the assignee of the rocking chair patent cannot practice the invention without a license from the holder of the chair patent without infringing the chair patent. Also, the assignee of the chair patent cannot practice the rocking chair invention without a license from the assignee of the rocking chair patent. Failure to do this would also result in infringement.

Now suppose that an inventor wanted to design a new chair and did not want to take a license from the assignee of the chair patent. To do this, the inventor would need to avoid a claim that had all of the elements of the chair patent. Such an effort is called a *design-around*. The inventor might create a chair with elements seat, legs, and lower retaining ring providing stability to the legs, but no back.

This new chair design is patentable over the first chair, because the patentability analysis speaks to the set of elements in the chair compared to all prior art. The new chair might, however, still infringe on another patent. How could this be? The relevant infringement analysis would examine the elements and might discover that the lower retaining ring was an article individually protected by patent, and thus the new device is patentable but cannot be practiced without first obtaining a license from the assignee of the lower retaining ring patent.

When a Patent Is in Your Way

Suppose you've got a great idea and you're ready to start making money with it. But somebody has a patent that seems uncomfortably close to what you want to do. What are your options? I give you five:

- Play it safe, quick, and possibly expensive by simply and straightforwardly taking a license to the patent.
- Wait until the patent expires and then produce your article and offer it for sale.
- Disregard the patent and wade through the murky and dangerous swamp of risking infringement. If you get stuck, a treble judgment might be rendered against you for not only infringing someone else's patent but for doing it willfully. You will also be estopped from further practicing your infringement.
- Pursue a design-around invention and come up with something close, but not too close, to the scope of art encompassed by the patent. This can be a rocky road, and it is often time-consuming and expensive. Also, there is no guarantee of ultimate success, and you might still be infringing the patent if the result of your design-around is found to be equivalent to the patent.
- "Take a chainsaw to the existing patent" in an attempt to have the patent, or some parts of it, declared void. This process is called *reexamination* and it is done by bringing the USPTO's attention (along with a fee, of course) to the prior art in the form of extant patents or publications that were overlooked during prosecution. There is a significant chance, historically about two-thirds, that a patent will fall or be narrowed in scope upon reexamination, so it is something to consider. It has also been a favored offense against so-called "patent trolls," individuals or companies that acquire assertable rights to a patent so that they can demand royalties.

Types of Patents

So far, we have looked at *utility* patents that protect the intellectual property covering how an article works and how it is used. But other patents protect how an article *looks*, and these patents are called *design* patents.

NOTE *There are also plant patents. Rose hybridization is a big business, for example.*

Often both a utility patent and a design patent will be sought for an article, particularly if the article is highly exposed to the public when it is in use. As our example, let's take a look at utility patent U.S. Patent 5814968, "Battery Charger and Rechargeable Electronic Paging Device Assembly." This item is described as "a battery charger and electronic device assembly for recharging stackable electronic devices in multiple orientation or rotation with respect to each other without the need for either removing batteries or plugging devices into charging racks."

These devices are often packaged as a drink coaster with lights that flash to notify a waiting customer that a table is ready. This requires that one device be handed out to each waiting customer party. In a popular restaurant at dinnertime, a queue of parties may be waiting for a table, and the restaurant will need a significant number of these devices on hand. However, as the patent's background section explains, or motivates, this requires that the service provider

> *either continually refresh the charge on the coaster device to prevent failure or to purchase a sufficient number of individual coasters to meet the maximum demand anticipated. Use of standard battery charger and rack assemblies during customer service hours, however, requires the provider to routinely shift their attention from the waiting customer to the charger assembly, to facilitate proper placement and orientation of the rechargeable coaster device in the charger module or rack.*

So what Patent 5814968 teaches is a way to build a paging coaster so that the charger can handle many coasters simultaneously; they can be stacked while charging and therefore take up only a nominal amount of space.

The patent's claims protect a battery charger and rechargeable device assembly (the coaster). Key to the claims are the exposed electrical terminals below and above each coaster that allow them to be stacked and recharged simultaneously.

But what might this look like? I suspect you've seen one. You might not recall it from the technical description, but let's look at the design patent, U.S. Des. 371054. Design patents have only one claim, and this one reads, "The ornamental design for a combined drink coaster and pager, as shown." The patent's figure 2 is shown in our Figure A-5. Now you know what those metal pieces in the corners are for!

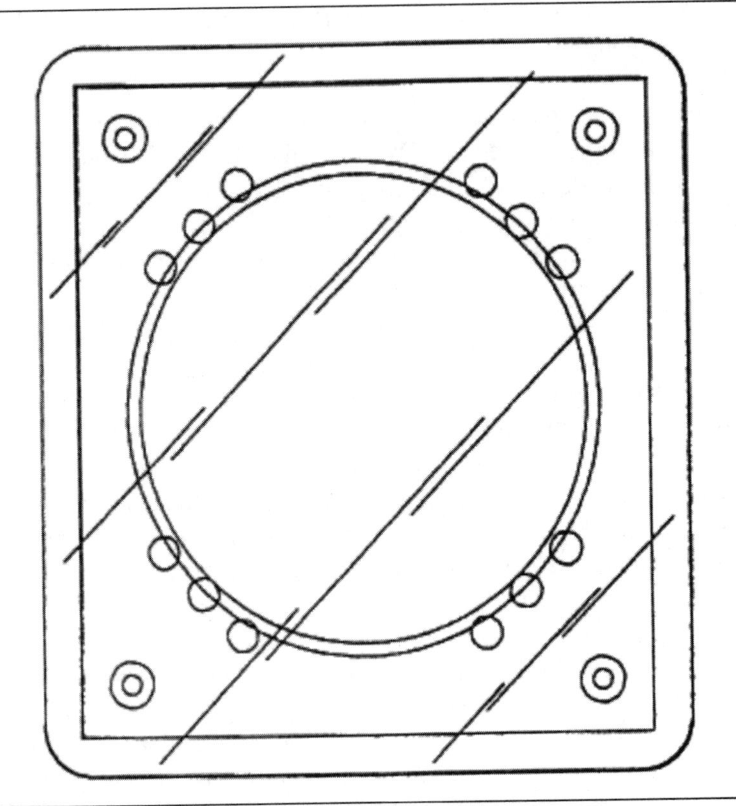

FIGURE A-5 Ornamental design for drink and coaster pager

Le Mot Juste

The right word. The ability to express yourself clearly, completely, and unambiguously is critical in working in the patent arts. A patent must teach someone of ordinary skill in the art (known as a "legal fiction" of patent law, just as a "reasonable person" is a legal fiction of tort law) in the art how to practice the invention. The patent does this using words and usually (but not always) diagrams. It is therefore essential that the patent employ clear and sufficient wording.

Frankly, this is difficult for many folks. Effective writing is not easily accomplished. The English language is rich and replete with so many candidates for its different parts of speech that picking just the right words can be a daunting task.

Consider the prepositions, for example. They are considered by some to form the most difficult class of words to teach. Their individual meanings, as words considered "in a vacuum," so to speak, are well defined but their employment is often idiomatically governed: You get *in* the car, but you get *on* the bus. And, there are so many prepositions: aboard, about, above, across, after, against, along, amid, amidst, among, amongst, around, aside, astraddle, astride, at, athwart, atop. And those are just a few of the prepositions beginning with the letter *a*.

On August 2, 1988, U.S. Patent 4761290, "Process for making dough products," was issued. The patent abstract is a pleasure to read if you're into carbs:

> *Dough products convertible by heating to light, flaky, crispy dough products are produced by applying shortening flakes to a dough, coating a light batter upon the dough product and heating the batter-coated dough to first set the batter and then subsequently melt the shortening flakes, thereby forming air cells within the batter and at the surface of the dough.*

Two of this patent's claims were to become infamous. As my point can be made with either of these two claims, I quote only the first, claim 1, and I have bolded a single word, a two-letter preposition:

> *1. A process for producing a dough product which is convertible upon finish cooking by baking or exposure to microwaves in the presence of a microwave susceptor into a cooked dough product having a light, flaky, crispy texture, which comprises the steps of:*
>
> *providing a dough;*
>
> *applying a layer of shortening flakes to at least one side of said dough;*
>
> *coating a light batter to a thickness in the range of about 0.001 inch to 0.125 inch over said at least one side of said dough to which said shortening flakes have been applied;*
>
> *heating the resulting batter-coated dough* **to** *a temperature in the range of about 400° F. to 850° F. for a period of time ranging from about 10 seconds to 5 minutes to first set said batter and then subsequently melt said shortening flakes, whereby air cells are formed in said batter and the surface of said dough; and*
>
> *cooling the resulting dough product.*

Do you see the problem? It was because of that choice of a two-letter preposition that we had the decision from the Court of Appeals for the Federal Circuit CAFC 03-1279, Chef America, Inc., v Lamb-Weston, Inc. The appeals hearing upheld a district court's finding of noninfringement. The essentials of the harangue can be quickly garnered by CAFC Judge Friedman's opening paragraph:

The sole issue in this appeal is the meaning of the following language in a patent claim: "heating the resulting batter-coated dough to a temperature in the range of about 400° F. to 850° F." The question is whether the dough itself is to be heated to that temperature (as the district court held), or whether the claim only specifies the temperature at which the dough is to be heated, i.e., the temperature of the oven (as the appellant contends). We agree with the district court that the claim means what it says (the dough is to be heated "to" the designated temperature range) and therefore affirm.

NOTE

The Court of Appeals for the Federal Circuit (CAFC) is a relatively recently established court system (1982) that handles patent disputes that are not resolved in the USPTO or at the District Court level. It is a stepping-stone to the Supreme Court. About a third of its cases have to do with patent disputes.

The CAFC further noted the following:

The claim requires "heating the resulting batter-coated dough to a temperature in the range of about 400° F. to 850° F." These are ordinary, simple English words whose meaning is clear and unquestionable. There is no indication that their use in this particular conjunction changes their meaning. They mean exactly what they say. The dough is to be heated to the specified temperature. Nothing even remotely suggests that what is to be heated is not the dough but the air inside the oven in which the heating takes place. Indeed, the claim does not even refer to an oven.

The problem is that if the batter-coated dough is heated to a temperature range of 400° F. to 850° F., as the claim instructs, it would be burned to a crisp. Instead of the "dough products suitable for freezing and finish cooking to a light, flaky, crispy texture," '290 patent, col. 2, ll. 11-12, which the patented process is intended to provide, the resultant product of such heating will be something that, in the words of one of the attorneys in this case, resembles a charcoal briquet. To avoid this result and to insure that the patented process can accomplish its stated objective, Chef America urges us to interpret the claim as if it read "heating the…dough at a temperature in the range of," i.e., to apply the heating requirement to the place where the heating takes place (the oven) rather than the item being heated (the dough).

This court, however, repeatedly and consistently has recognized that courts may not redraft claims, whether to make them operable or to sustain their validity.

A nonsensical result does not require the court to redraft the claims of the ['290] patent. Rather, where as here, claims are susceptible to only one reasonable interpretation and that interpretation results in a nonsensical construction of the claim as a whole, the claim must be invalidated.

Then, citing a decision from another case, the CAFC continued:

Where, as here, the claim is susceptible to only one reasonable construction, the canons of claim construction cited by [Chef America] are inapposite, and we must construe the claims based on the patentee's version of the claim as he himself drafted it.

Thus, in accord with our settled practice we construe the claim as written, not as the patentees wish they had written it. As written, the claim unambiguously requires that the dough be heated to a temperature range of 400° F. to 850° F.

Although "a patentee can act as his own lexicographer to specifically define terms of a claim contrary to their ordinary meaning" [again, citing another case], we discern nothing in the claims, the specification, or the prosecution history that indicates that the patentees here defined "to" to mean "at." To the contrary, the prosecution history suggests that the patentees intentionally used "to" rather than "at" in drafting the temperature requirements of the claim.

Finally, the lesson is succinctly stated in the position of the district court's opinion: "It is the job of the patentee, and not the court, to write patents carefully and consistently." (We note that on June 29, 2004, the USPTO issued a certificate of correction to U.S. Patent 4761290 replacing the "to" with "at" in the claims.)

Why Bother with Dependent Claims?

You know that dependent claims fall within the deed or scope of the independent claim from which they depend. Now you might be asking, Isn't anything claimed via a dependent claim also implicitly claimed by the independent claim? What then is the need for dependent claims?

It turns out that dependent claims serve a variety of uses. Perhaps the most important use is in drafting the patent application for prosecution. Suppose a piece of prior art had escaped notice and was turned up by the patent examiner. If the independent claim were all you had, the claim would be disallowed and you would be stuck, sadder but wiser. But if you had included one or more narrowing dependent claims, you could fold the independent claim into a dependent claim and proceed on. Graphically, the process looks like the drawing in Figure A-6.

Another reason for using dependent claims is to capture much of the improvements and useful refinements to the subject of the independent claim. Remember the rocking chair? It was certainly embraced by the patent for the chair, but it was separately patentable, although it should not be practiced without a license from the holder of the chair patent—but neither can the chair patent holder practice the rocking chair without a license from its holder. The message is simple: wrap up all you can or someone else might.

A third reason for using independent claims is concerned with infringement prosecution. When your patent is being explained to a jury, whose constituents are probably not folks of ordinary skill in the art, it might be difficult to convince them that the infringer's widget actually infringes your first claim. Even though the words legally encompass your product, they might be far too abstruse for the average juror. But if your subsequent claims go into great detail as to how your widget is constructed, your attorney can clearly point to claim element after claim element, bringing them to life as your widget is shown to embody them exactly as the patent unfolds. In such a situation, your patent can be a veritable roadmap to protecting the fruits of your labor.

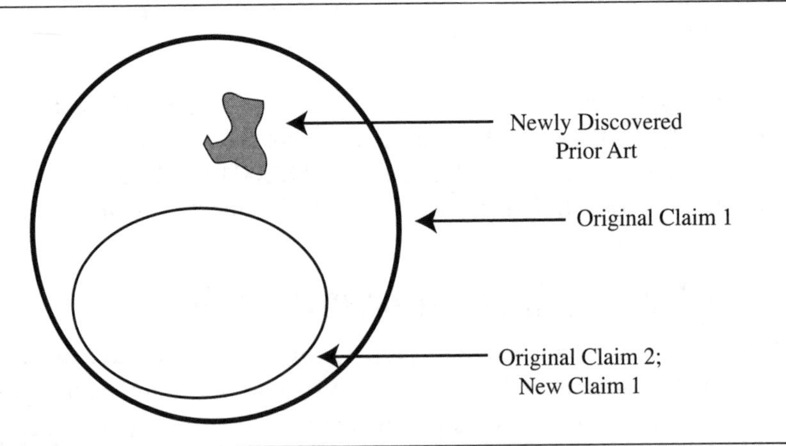

FIGURE A-6 A gambit for necessary narrowing of scope

Speed Is of the Essence and Will Become More So

Currently, the United States is in a distinct minority when it comes to patent priority. The rest of the world accords candidacy for patent issuance to the first inventor to file an application. Not so the United States. As of this writing, the United States allows the first-to-invent inventor to have candidacy for patent issuance providing that the first-to-invent inventor meets a host of requirements involving secrecy and diligence. This will change, however, when the United States moves to bring its patent system into comportment with other nations. When that happens, and it will probably be very soon, speed will be of the essence.

U.S. Patent application US20030052226, "Aircraft integrated non-lethal weapon system," was filed on September 19, 2001, just eight days after the attack of 9/11. Here is the first sentence of the background section: "The tragic events of September 11, 2001, underscores dramatically the need for preventing ingress of hijackers into the cockpits of commercial airplanes." Now that's speed!

Patentability: §101 Revisited

One of the categories of statutorily unpatentable items includes items found in nature, such as wild plants, gravitation, and naturally occurring elements such as iron and oxygen. In 1952, Congress passed a major piece of legislation affecting the U.S. patent process. In this law is a famous phrase respecting candidates for patentability: "anything under the sun that is made by man." Although it predated this phrase by a few years, U.S. Patent 3156523, "Element 95 and method of producing said element," is a very interesting case in point. This patent was assigned to the United States by its inventor, Dr. Glenn T. Seaborg, a true national luminary, who received the Nobel Prize in Chemistry in 1951. Dr. Seaborg was a most accomplished scientist and was the author of scores of books, a revered professor, and head of the Atomic Energy Commission.

What makes this patent of Dr. Seaborg's so interesting is that it concerns the element 95, Americium. Americium is a manmade element and was, evidently therefore, considered eligible for patenting. The first claim of the '523 patent is possibly the shortest claim ever allowed. It reads simply, "1. Element 95."

Americium, which immediately follows plutonium in the periodic table, is a member of the actinide series of metallic, radioactive, and chemically similar metals. It is used in some smoke detectors.

Sometimes It's a Secret

When a patent application is submitted to the USPTO, one of the first actions is a review of the application by representatives of the national authorities. Occasionally these authorities deem that public disclosure of the material within the application would be prejudicial to the security interests of the United States. This might lead to a secrecy order and keep the application "on ice" until publicly known technology has sufficiently progressed to render the suppressed disclosure harmless.

I think the record for the longest such suppression must lie with U.S. Patent 6097812, filed July 25, 1933, and issued August 1, 2000. The patent was for a cryptographic system and the inventor was William F. Friedman, who was head of research for the Army's Signals Intelligence Service. Friedman died in 1969. His patent is assigned to the United States of America as represented by the National Security Agency.

Claim All of It

Sometimes inventors do not fully appreciate what they have invented. And sometimes potentially valuable protection is not seized from the start. The reason for this could be that the inventor is so focused on invention and patenting a system that some of the less glamorous system components or methods are invented and ignored as mere stepping stones to get to the prize. The good news is that there is a way to correct such an oversight. The bad news is that a certain amount of adverse exposure is still assumed by such an oversight.

A classic example of not recognizing all that had been invented is in the patent pair U.S. 5119104 "Location System Adapted for Use in Multipath Environments," issued June 2, 1992, and its reissue Re. 36,791, reissued July 25, 2000. In the first of these patents, the inventor envisioned a local area network (LAN)-based radiolocation system for tracking objects equipped with a "TAG" transmitter. The patent's abstract is shown next along with the patent's figure 1b placed in our Figure A-7.

A radiolocation system for multipath environments, such as for tracking objects in a semiconductor fabrication facility [10] (FIGS. 1a-1b), includes an array of receivers (20) distributed within the tracking area, coupled to a system processor (40) over a LAN. A TAG transmitter (30) located with each object transmits, at selected intervals, spread spectrum TAG transmissions including at least a unique TAG ID. In a high resolution embodiment, object location

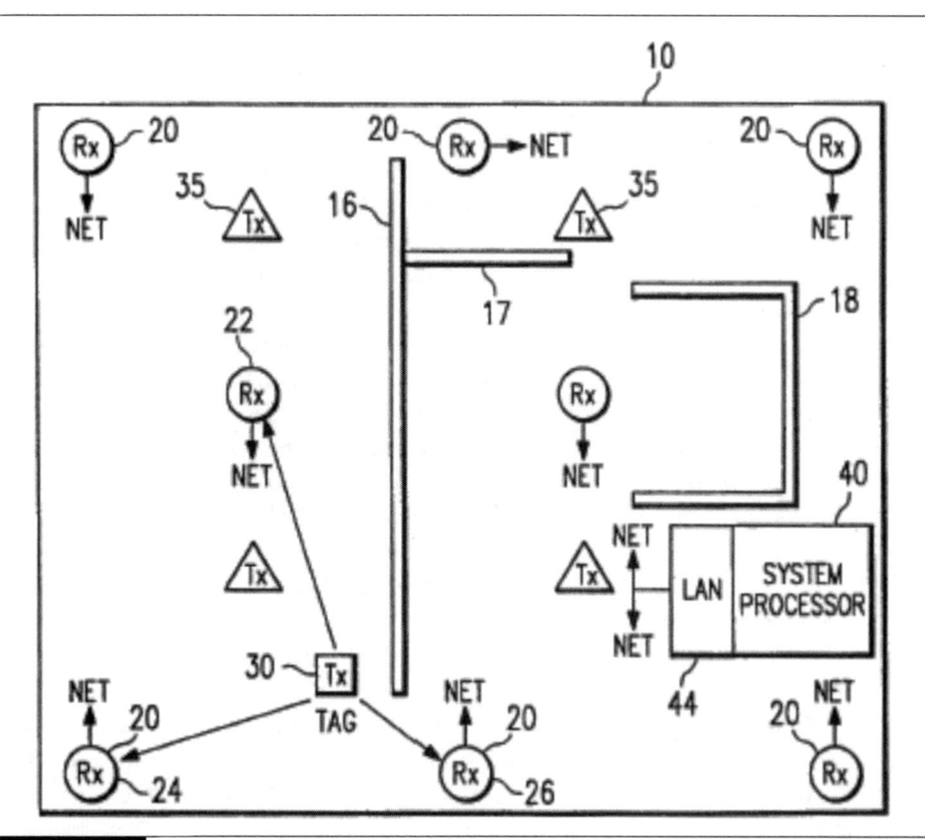

FIGURE A-7 Patent figure 1b

> *is accomplished by time-of-arrival (TOA) differentiation, with each receiver (FIG. 2b) including a TOA trigger circuit (64) for triggering on arrival of a TAG transmission, and a time base latching circuit (65) for latching the TOA count from an 800 MHz time base counter. In a low resolution embodiment, each receiver of the array is assigned a specific location-area, and receives TAG transmissions almost exclusively from TAGs located in that area, thereby eliminating the need for any time-of-arrival circuitry."*

As described, the invention performs the utile function of locating TAGs with transmitters so that they can be tracked within an operational area. The claims speak to this. Let's look at claim 1:

A location system for locating objects within a tracking environment using time-of-arrival differentiation for electromagnetic transmissions received at multiple receivers, comprising:

for each object, a TAG transmitter for transmitting, at selected intervals, TAG transmissions that include a unique TAG ID;

an array of receivers distributed within the tracking environment such that a TAG transmission is received by at least three receivers;

each receiver including a time-of-arrival circuit and a data communications controller;

the time-of-arrival circuit is responsive to the arrival of a TAG transmission for providing a TOA count corresponding to the time-of-arrival of the most direct path for such TAG transmission, with the TOA count being synchronized to a system synchronization clock provided to each receiver;

the data communications controller is responsive to the receipt of a TAG transmission for providing a corresponding TOA-detection packet that includes the associated TAG ID and TOA count; and

a location processor for receiving the TOA detection packets, and for determining the location of each TAG, and its associated object, from at least three corresponding TOA-detection packets received from different receivers.

So, according to claim 1, a TAG transmits "at selected intervals." Is that all there is to initiating TAG transmissions, or is there more? Let's look at what's taught in the specification. At our point of interest, the inventor teaches according to the patent's figure 2a shown in our Figure A-8.

In the patent, we find:

To conserve power and to increase the available population of TAG transmitters, each spread spectrum transmitter 52 is normally in a power-saver mode, being enabled for transmission by battery saving circuit 54 only while its associated object is being moved to a new location. Object motion is detected by motion detector 56, which provides an appropriate indication to the battery saving circuit.

In response to a motion indication, battery saving circuit 54 initiates a transmit mode by enabling spread spectrum transmitter 52 for an initial TAG transmission. The TX-packet in this initial TAG transmission includes, in addition to the TAG ID, a Motion Initiated status.

While the object remains in motion (as detected by motion detector 56), periodicity control 58 causes spread spectrum transmitter 52 to re-transmit

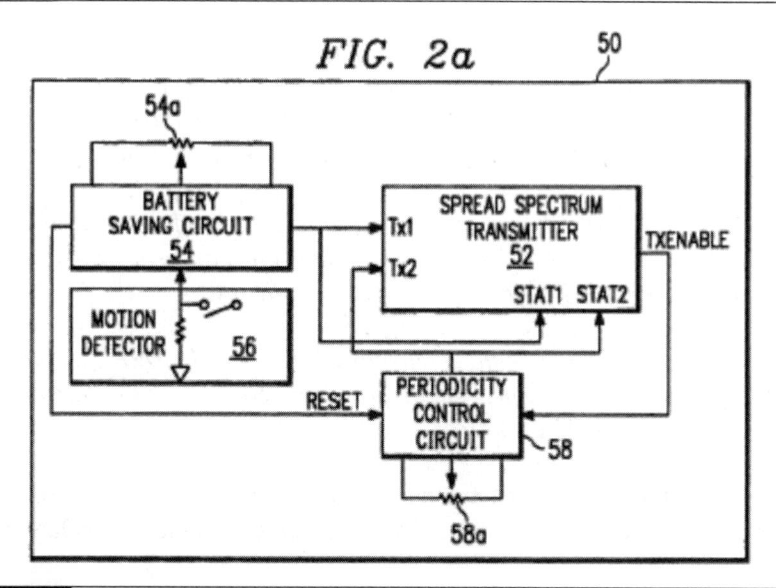

FIGURE A-8 Patent's figure 2a

TAG transmissions at selected intervals (such as every 15 seconds). The TX-packets in these periodic TAG re-transmissions include, in addition to the TAG ID, a Motion Continuing status.

When the object arrives at its new location and becomes stationary, motion detector 56 stops providing an object motion indication to battery saving circuit 54. After a predetermined period in which the object is stationary (such as 30 seconds), the battery saving circuit disables periodicity control 58, and causes the spread spectrum transmitter to transmit a final TAG transmission with a TX-packet including a Motion Stopped status.

Coupling a motion detector with a transmitter is a neat idea, but it wasn't claimed. Could this still be a patentable teaching? The inventor came to believe so and sought protection through what is termed a "broadening reissue." An application for a broadening reissue must be filed within two years of the date of issue of the original patent, and this was done on June 2, 1994, two years to the day.

In application seeking reissue, the applicant must fulfill Patent Rule 1.175(a), which requires:

(a) The reissue oath or declaration ... must also state that:

(1) The applicant believes the original patent to be wholly or partly inoperative or invalid by reason of a defective specification or drawing, or by reason of the patentee claiming more or less than the patentee had the right to claim in the patent, stating at least one error being relied upon as the basis for reissue; and

(2) All errors being corrected in the reissue application up to the time of filing of the oath or declaration under this paragraph arose without any deceptive intention on the part of the applicant.

NOTE

This rule was changed effective December 1, 1997. The old rule required the patentee to provide details about each error's nature and how the errors came about.

Going through the first amended reissue application declaration in the file wrapper for Re. 36,791 we find that the patentee did indeed declare in part:

I believe the original patent to be wholly or partly inoperative or invalid by reason that I claimed less than I had a right to claim in the patent—that is, I failed to claim novel features of the invention broadly enough. The reason why I believe this is because in my patent, I did not broadly claim a transmitter alone that transmits the response to motion and periodically transmits in response to a lack of motion.

The reissue application was eventually successful and issued as U.S. Patent Re. 36,791. The first of the newly allowed claims, claim 55, reads as follows:

A transmitter including:

transmitter circuitry for transmitting information; and

a motion detection circuit for detecting motion and lack of motion of the transmitter, said motion detection circuit enabling the transmit circuitry to transmit information at first selected intervals when the transmitter is in motion and enabling the transmit circuitry to transmit information periodically at second selected intervals in response to detecting lack of motion,

wherein said second selected intervals are at a low duty cycle relative to said first selected intervals.

To avoid not claiming it all, to be more assured that the whole invention is presented to the USPTO for protection, it is probably a best practice to invite

FIGURE A-9 A suggested review process

company colleagues to review your invention disclosure, as sketched in Figure A-9. An inventor is often too close to the invention to see valuable aspects that he or she might have considered obvious. Remember that you, the inventor, are at the center of the particular maelstrom of your invention. A 30,000 foot view cannot hurt and might protect the assignee's—your company's—best interests.

Recap

Patenting is only one of the measures that can be applied to protect your inventions. If protecting your invention from arrogation and exploitation by others is a concern, you should seek legal advice to determine the best route to accomplish your objectives.

If you do decide to pursue a patent on your invention, you may represent yourself but I suggest you meet with a patent attorney or a registered patent agent. A roster of patent attorneys and patent agents is maintained for public use at https://oedci.uspto.gov/OEDCI/.

Before you meet with a patent attorney or registered patent agent, have your invention clearly defined, know how to implement and practice it, and consider how it might be broadened and futurized, as discussed in Chapter 1. Remember that your Eureka! moment might have produced more invention than you realize. Some additional material that might have seemed "obvious" to you may be non-obvious to someone of ordinary skill in the art, and you might want to broaden the scope of your application to include this material.

When reviewing the patent application prepared for you, read it carefully and be certain that the claims clearly and carefully describe what you understand to be your invention, and ensure that that you have taught, but not necessarily specifically identified, the best mode for implementing it. You should also review the question of inventorship as discussed in Appendix C.

Discussions and Reflections

- A patent can be like fence built around our intellectual property. The strength of the fence is in the preparation of the patent. The specification must teach appropriately and the claims must be strong and carefully tailored in their successive differentiation from broad to specific.

  To prove literal infringement, every element of a claim must be present in the inventor's device for the device to "read onto" a patent's claim and for there to be literal infringement. (Recall that this is the "all elements rule.")

  The construction of claims, then, starting with an independent claim and proceeding through the associated dependent claims, must successfully and tightly cover the invention to be protected. Starting with a claim that is too restrictive might render the patent weak in carving out white space.

  Consider U.S. Patent 5587701, "Portable alarm system." Its abstract reads as follows:

  A portable alarm system is disclosed in which the alarm functions are contained within a portable enclosure, communication is maintained between the enclosure and wireless security contacts placed at points of entry, and the alarm is capable of initiating a telephone call to a security monitor station either by conventional hard wired telephone lines within a building, or by cellular transmission, or via 800 MHz trunking.

  Following is claim 1. Analyze it and discuss your assessment. Pay particular attention to the sixth element of the claim, the audio siren.

  An alarm device, comprising:

  a single portable enclosure, said enclosure including a handle to enable a person to carry said enclosure by hand;

  a user interface control panel secured within said enclosure, wherein said interface control panel is accessible to a user of said alarm device;

a microprocessor board installed within said enclosure, said board in communication with said interface control panel;

a signal receiver installed within said enclosure, said receiver in communication with said microprocessor board, and wherein said receiver receives signals from at least one zone within a structure being monitored;

a communication circuit secured within said enclosure, and independent of any hard wired telephone lines connected to said structure, said communication circuit adapted to initiate a telephone call to a location apart from said structure;

an audio siren connected to said microprocessor board, wherein said siren is electrically connected to said microprocessor board so as to sound when a signal is received at said signal receiver indicating said zone of said structure has been breached; and

a data interface electrically connected to said communication circuit and said microprocessor board for communicating a signal to said communication circuit from said microprocessor board to cause said communication circuit to initiate a telephone call.

■ The USPTO likes drawings and in its rules we find the following:

- The applicant for a patent is required to furnish a drawing of his or her invention where necessary for the understanding of the subject matter sought to be patented;
- Whenever the nature of the subject matter sought to be patented admits of illustration by a drawing without its being necessary for the understanding of the subject matter and the applicant has not furnished such a drawing, the examiner will require its submission.

U.S. Patent 4761290, "Process for making dough products," has no diagrams. Try to construct a diagram that will make the desired "at" interpretation clear or at least call the drafter's attention to the potential ambiguity.

■ The identification of the correct inventorship of a patent is a critical issue, because recordation of a faulty inventorship can render the patent invalid. But what constitutes inventorship? The answer is very simple and intimately involves the patent's claims. Someone is an inventor if and only if he or she has significantly contributed to at least one claim. For quick mental sport, consider that U.S. Patent 6434583 cites 29 inventors and has

11 claims. What is the maximum value of x that guarantees the following sentence to be true? "One claim of U.S. Patent 6434583 must have had contributions from at least x inventors."

A Short Personal Story Respecting Inventorship

We'll delve into inventorship in more detail in Appendix C. For now, I'll tell a quick war story to reinforce the point. Inventorship devolves from contribution, not from rank, position, or friendship. Some cultures are prone to including high-ranking company officials into the inventorship by virtue of their status.

A few years ago, I was on a mergers and acquisitions (M&A) due diligence team and I asked to see the company's patent portfolio. The files were trotted out and the company's intellectual property counsel joined the group. The CEO sat behind his desk as I slowly paged through the documents. He seemed to be a very pleasant fellow, and he definitely had more of a business bent than engineering, hardly a surprise. But on picking up yet another patent, there it was. The inventors included the CEO in the patent, and I decided to test the waters. "This is very impressive," I said to the CEO. "I see you are one of the inventors. What was your contribution to this?" A broad smile spread across his face and then he said, "Well, that's a good question." And then, faster than you can say "inventorship," the IP counsel boomed the assurance that it was a very important contribution. Good to know.

Appendix B

U.S. Patent 6004596 "Sealed Crustless Sandwich"

United States Patent [19]

Kretchman et al.

[11] **Patent Number:** **6,004,596**

[45] **Date of Patent:** **Dec. 21, 1999**

[54] **SEALED CRUSTLESS SANDWICH**

[75] Inventors: **Len C. Kretchman**, Fergus Falls, Minn.; **David Geske**, Fargo, N. Dak.

[73] Assignee: **Menusaver, Inc.**, Orrville, Ohio

[21] Appl. No.: **08/986,581**

[22] Filed: **Dec. 8, 1997**

[51] **Int. Cl.**^{6} **A21D 13/00**

[52] **U.S. Cl.** **426/94**; 426/274; 426/275; 426/297

[58] **Field of Search** 426/94, 274, 275, 426/297, 138

[56] **References Cited**

U.S. PATENT DOCUMENTS

3,083,651	4/1963	Cooper	426/275
3,690,898	9/1972	Partyka	426/275
3,767,823	10/1973	Wheeler et al.	426/275
3,769,035	10/1973	Kleiner et al.	426/275
3,862,344	1/1975	Zobel	426/244
4,382,768	5/1983	Lifshitz et al.	426/275
5,853,778	12/1998	Mayfield	426/89

OTHER PUBLICATIONS

"50 Great Sandwiches", Carole Handslip, pp. 81–84,86,95, 1994.

Primary Examiner—Lien Tran
Attorney, Agent, or Firm—Vickers, Daniels & Young

[57] **ABSTRACT**

A sealed crustless sandwich for providing a convenient sandwich without an outer crust which can be stored for long periods of time without a central filling from leaking outwardly. The sandwich includes a lower bread portion, an upper bread portion, an upper filling and a lower filling between the lower and upper bread portions, a center filling sealed between the upper and lower fillings, and a crimped edge along an outer perimeter of the bread portions for sealing the fillings therebetween. The upper and lower fillings are preferably comprised of peanut butter and the center filling is comprised of at least jelly. The center filling is prevented from radiating outwardly into and through the bread portions from the surrounding peanut butter.

10 Claims, 4 Drawing Sheets

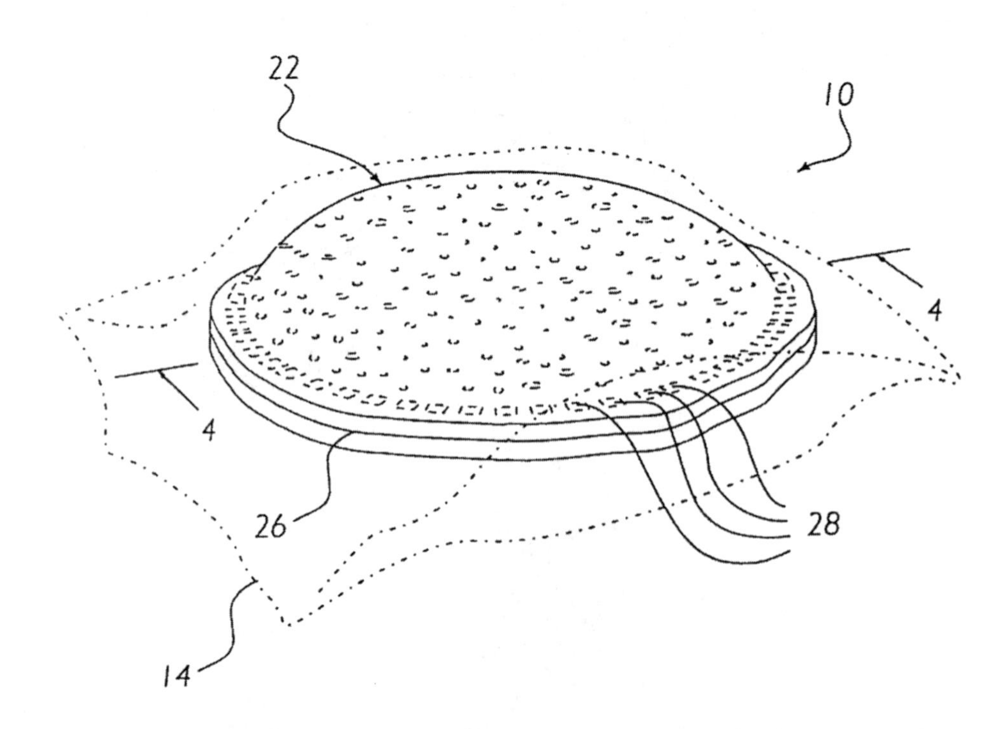

U.S. Patent Dec. 21, 1999 Sheet 1 of 4 **6,004,596**

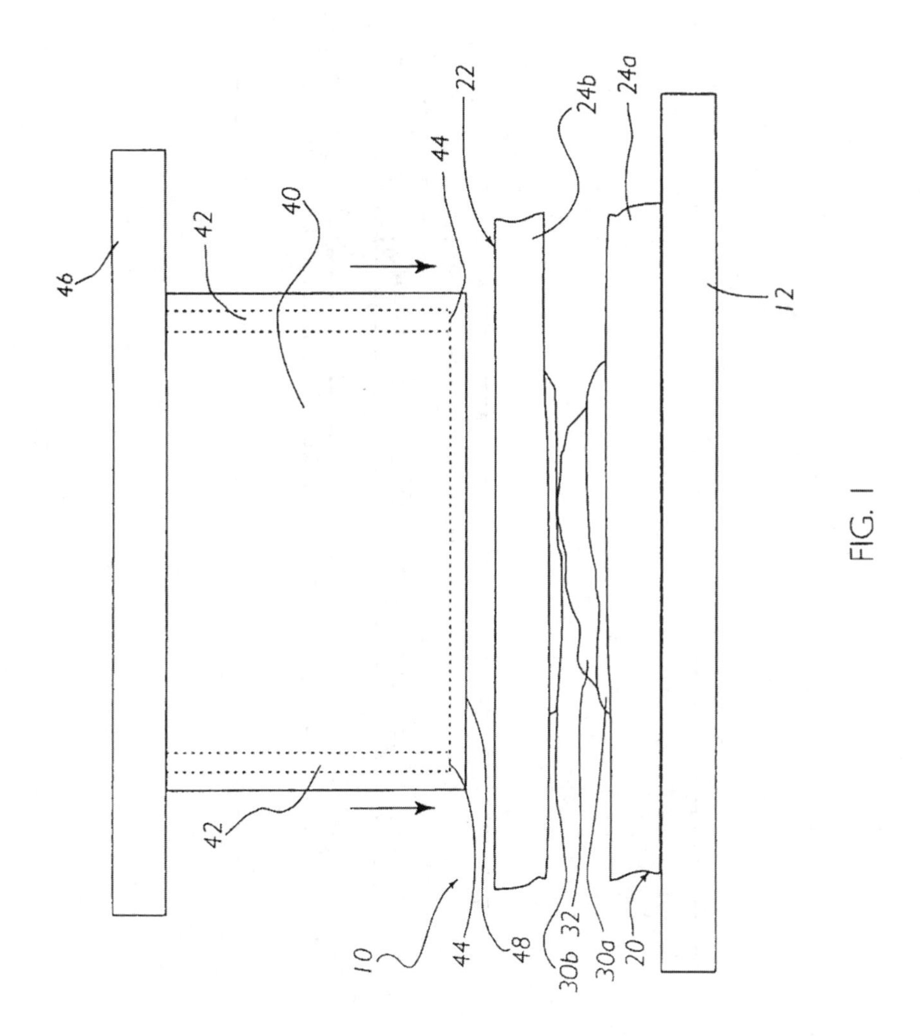

FIG. 1

U.S. Patent Dec. 21, 1999 Sheet 2 of 4 6,004,596

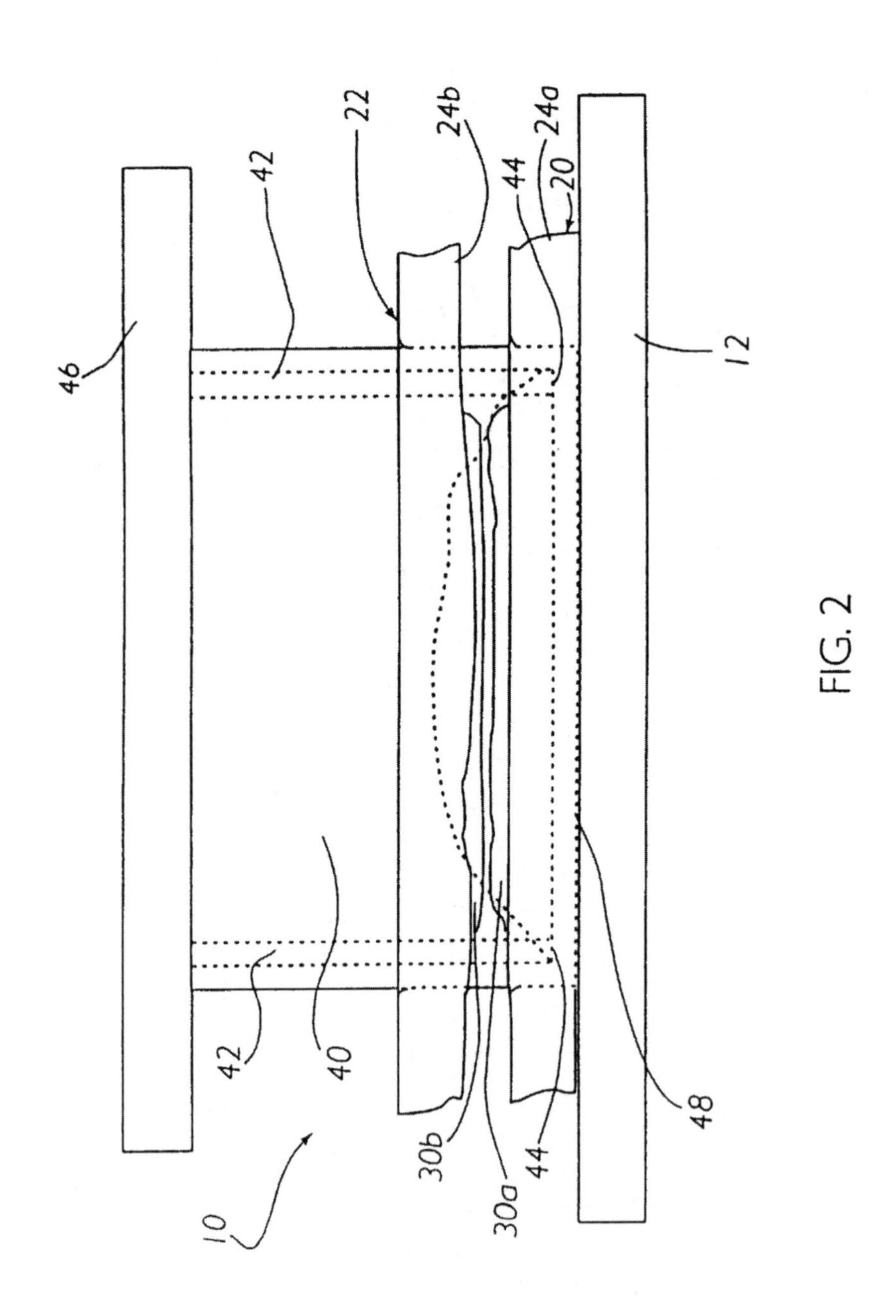

FIG. 2

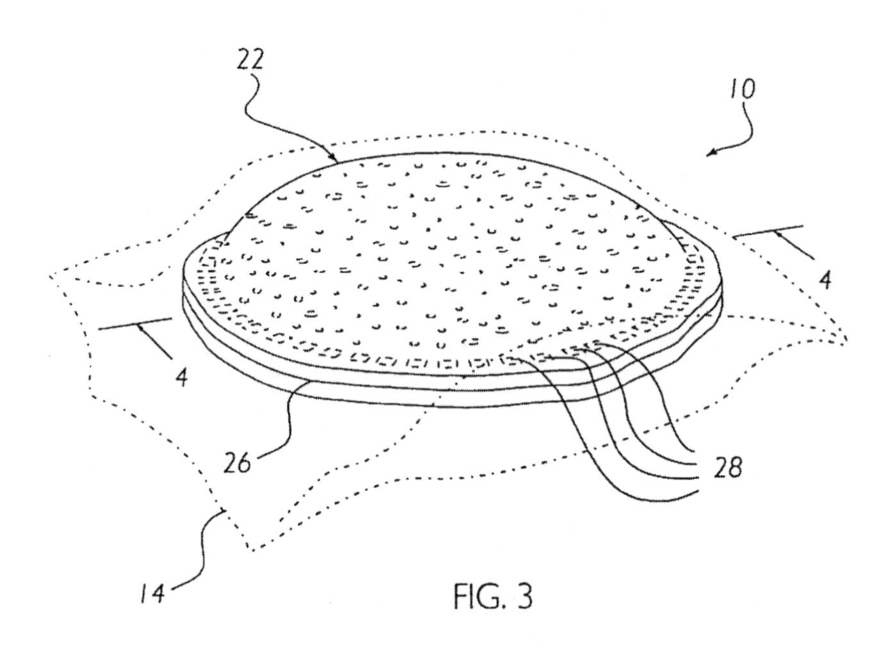

FIG. 3

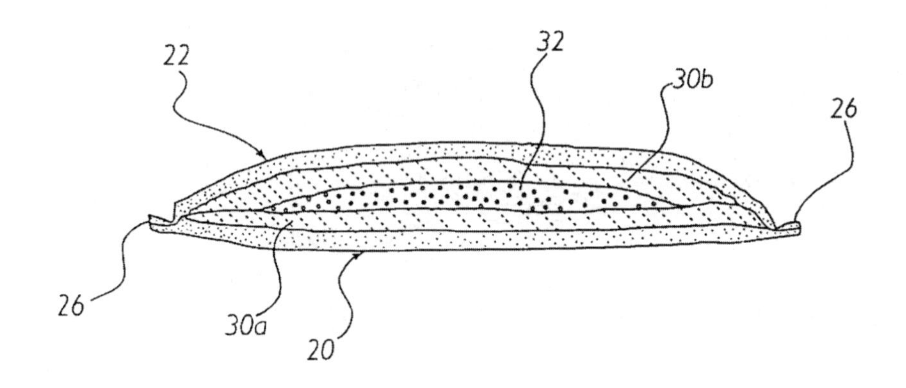

FIG. 4

U.S. Patent Dec. 21, 1999 Sheet 4 of 4 **6,004,596**

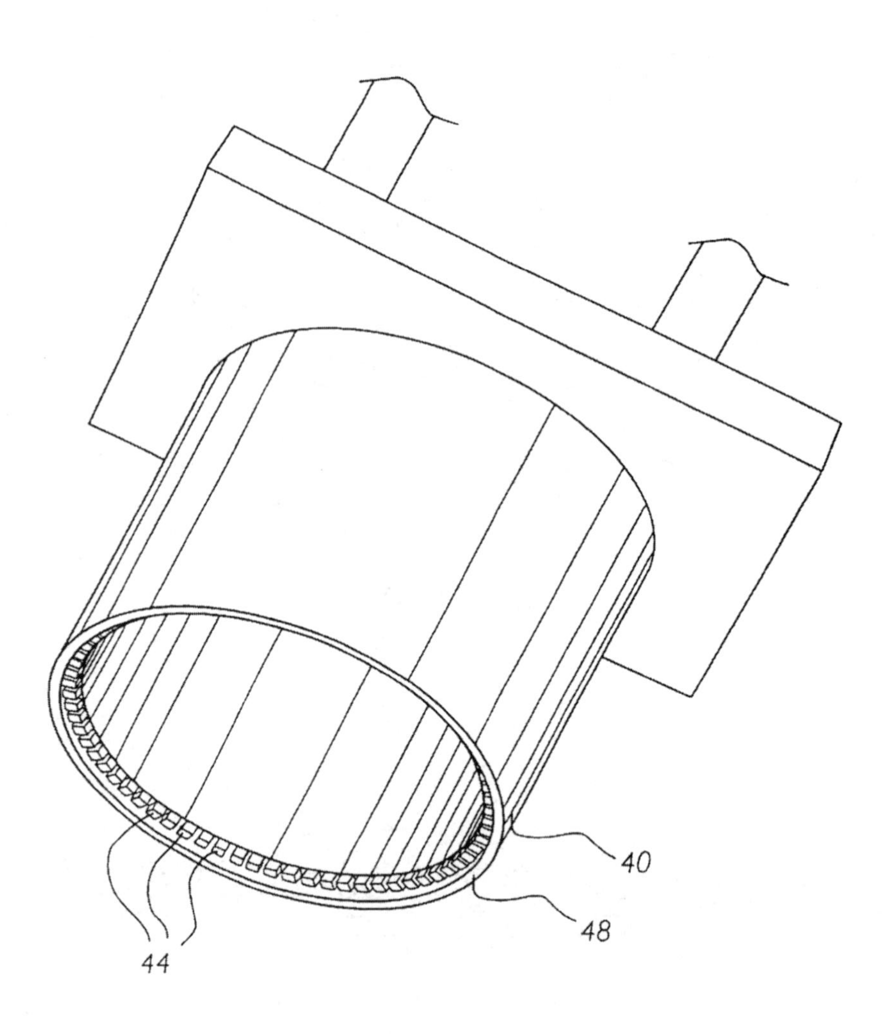

FIG. 5

SEALED CRUSTLESS SANDWICH

BACKGROUND OF THE INVENTION

1. Field of the Invention

The present invention relates generally to sandwiches and more specifically it relates to a sealed crustless sandwich for providing a convenient sandwich without an outer crust which can be stored for long periods of time without a central filling from leaking outwardly.

Many individuals enjoy sandwiches with meat or jelly like fillings between two conventional slices of bread. However, some individuals do not enjoy the outer crust associated with the conventional slices of bread and therefore take the time to tear away the outer crust from the desired soft inner portions of the bread. This outer crust portion is then thrown away and wasted. There is currently no method or device for baking bread without having an outer crust. Hence, there is a need for a convenient sandwich which does not have an outer crust and which is not prone to waste of the edible outer crust portions. The present invention provides a method of making a sealed crustless sandwich which can be stored for extended periods of time without an inner filling from seeping into the bread portion.

2. Description of the Prior Art

There are numerous sandwich devices. For example, U.S. Pat. No. 3,690,898 to Partyka; U.S. Design Pat. No. 252,536 to Goglanian; U.S. Design Pat. No. 293,040 to Gagliardi; U.S. Design Pat. No. 317,672 to Presl; U.S. Design Pat. No. 318,360 to Sam; and U.S. Pat. No. 5,500,234 to Russo all of which are illustrative of such prior art.

While these sandwiches may be suitable for the particular purpose to which they address, they are not as suitable for providing a convenient sandwich without an outer crust which can be stored for long periods of time without a central filling from leaking outwardly. The prior art does not teach a sandwich without an outer crust which sealably retains an inner filling for extended periods of time.

In these respects, the sealed crustless sandwich according to the present invention substantially departs from the conventional concepts and designs of the prior art, and in so doing provides a sandwich primarily developed for the purpose of providing a convenient sandwich without an outer crust which can be stored for long periods of time without a central filling from leaking outwardly.

SUMMARY OF THE INVENTION

A primary object of the present invention is to provide a sealed crustless sandwich that will overcome the shortcomings of the prior art devices.

Another object is to provide a sealed crustless sandwich that does not have any crust.

An additional object is to provide a sealed crustless sandwich that retains an inner filling from seeping into the bread portion.

A further object is to provide a sealed crustless sandwich that can be stored for extended periods of time for use in lunch box type of situations.

Another object is to provide a sealed crustless sandwich that reduces the amount of wasted bread because of thrown away crust portions.

Another object is to provide a method of producing a sealed crustless sandwich.

Further objects of the invention will appear as the description proceeds.

To the accomplishment of the above and related objects, this invention may be embodied in the form illustrated in the accompanying drawings, attention being called to the fact, however, that the drawings are illustrative only, and that changes may be made in the specific construction illustrated and described within the scope of the appended claims.

BRIEF DESCRIPTION OF THE DRAWINGS

Various other objects, features and attendant advantages of the present invention will become fully appreciated as the same becomes better understood when considered in conjunction with the accompanying drawings, in which like reference characters designate the same or similar parts throughout the several views, and wherein:

FIG. **1** is a side view of cutting cylinder above the upper and lower bread with the fillings in between.

FIG. **2** is a side view of the cutting cylinder penetrating and crimping the upper and lower bread with the fillings in between.

FIG. **3** is an upper perspective view of the sealed crustless sandwich within an airtight packaging.

FIG. **4** is a cross sectional view from FIG. **3** disclosing the peanut butter sealing the jelly in between.

FIG. **5** is a lower perspective view of the cutting cylinder.

DESCRIPTION OF THE PREFERRED EMBODIMENT

Turning now descriptively to the drawings, in which similar reference characters denote similar elements throughout the several view, FIGS. **1** through **5** illustrate a sealed crustless sandwich **10**, which generally comprises lower bread portion **20**, an upper bread portion **22**, an upper filling **30***b* and a lower filing **30***a* between the lower bread portion **20** and upper bread portion **22**, a center fillng **32** sealed between the upper filling **30***b* and the lower filling **30***a*, and a crimped edge **26** along an outer perimeter of the bread portions **20**, **22** for sealing the fillings **30***a–b*, **32** therebetween. The upper filing **30***b* and the lower filling **30***a* are preferably comprised of peanut butter but may consist of any other edible substance such as but not limited to meat, vegetable oil, jelly, cheese, honey, or fruit. The center filling **32** is preferably comprised of jelly but may consist of any other edible substance such as but not limited to meat, vegetable oil, jelly, cheese, honey, or fruit. The center filling **32** is prevented from leaking outwardly into and through the bread portions **20**, **22** from the surrounding upper filling **30***b* and lower filling **30***a*. The sealed crustless sandwich is preferably packaged within a resilient packaging **14** to extend its useful life and for providing convenience for the user.

As shown in FIGS. **1**, **2** and **5** of the drawings, a cutting cylinder **40** has a sleeve **42** positioned within. The sleeve **42** preferably is sidably positioned within a lumen of the cutting cylinder **40**, but may be secured within the cutting cylinder **40**. The bottom edge of the sleeve **42** has a notched end **44** with the notches spaced approximately every ⅛ inch. The plurality of extensions created from the plurality of notches forms a corresponding plurality of depressions **28** in the crimped edge **26** which represent the pressure points where extensions have projected into the bread portions **20**, **22**. A cutting edge **48** of the cutting cylinder **40** is utilized to penetrate through the bread portions **20**, **22** as shown in FIG. **2** of the drawings. The cutting edge **48** may be formed into various shapes to form a unique design for the sealed crustless sandwich **10**. The notched edge of the sleeve **42** compresses the upper bread portion **22** into the lower bread portion **20** to form a seal which retains itself for extended periods of time. A support member **46** or similar structure is attached to the cutting cylinder **40** and the sleeve **42** as shown in FIGS. **1**, **2** and **5**, wherein the support member **46** is attached to an elevating/descending means for operating the cutting cylinder **40**.

As best shown in FIG. **4**, the upper filling **30*b*** is juxtaposed to a lower surface of the upper bread portion **22**. The lower filling **30*a*** is juxtaposed to an upper surface of the lower bread portion **20**. Preferably, the upper filling **30*b*** and the lower filling **30*a*** do not extend into the crimped edge **26** since any foreign substance within the crimped edge **26** weakens the seal between the lower and upper bread portions **20**, **22**. The center filling **32** is positioned and sealed between the upper filling **30*b*** and the lower filling **30*a*** as shown in FIG. **4** of the drawings. The crimped edge **26** preferably has a plurality of depressions **28** formed into from the pressure points caused by the notched end **44** of the sleeve **42**. The depressions **28** prevent the crimped edge **26** from separating thereby retaining the fillings **30*a–b***, **32** within.

In use, the upper surface of the lower bread portion **20** is partially covered with the lower filling **30*a*** over a defined area. The defined area is preferably inside of an inner perimeter of the sleeve **42** so as to eliminate or reduce the amount of lower filling **30*a*** within the crimped edge **26**. The center filling **32** is positioned centrally onto the lower filling **30*a*** as shown in FIG. **1** of the drawings. The lower surface of the upper bread portion **22** is partially covered with the upper filling **30*b*** over an area substantially equal to the defined area of the lower filling **30*a***. The upper bread portion **22** is positioned above the lower bread portion **20** with the upper filling **30*b*** juxtaposed to the center filling **32** and the lower filling **30*a***. The cutting cylinder **40** is descended onto the upper bread portion **22** as shown in FIG. **1** of the drawings. The cutting cylinder **40** penetrates the bread portions **20**, **22** so cut out circular portion surrounding the fillings **30*a–b***, **32**. The sleeve **42** is simultaneously descended onto the upper bread portion **22** whereby the notched end **44** engages the upper bread portion **22**. The notched end **44** forces the circular portion of the upper bread portion **22** onto the corresponding circular portion of the lower bread portion **20** thereafter crimping the bread portions **20**, **22** between the notched end **44** and a plate **12** supporting the lower bread portion **20** as shown in FIG. **2** of the drawings. Simultaneously during the crimping, the upper filling **30*b*** is forced into the lower filling **30*a*** surrounding the center filling **32**, thereby sealing the center filling **32** therebetween. The cutting cylinder **40** and the sleeve **42** are elevated away from the sealed crustless sandwich **10** while pressurized air is released into the cutting cylinder **40** to help force the sealed crustless sandwich **10** out from within. The crust **24** portion of the upper and lower bread portions **20**, **22** is hence removed from the sealed crustless sandwich **10** as shown in FIG. **2** of the drawings. After the sealed crustless sandwich **10** is removed from the cutting cylinder **40** and sleeve **42**, the air tight resilient packaging **14** is applied around it for preserving the sealed crustless sandwich.

As to a further discussion of the manner of usage and operation of the present invention, the same should be apparent from the above description. Accordingly, no further discussion relating to the manner of usage and operation will be provided.

With respect to the above description then, it is to be realized that the optimum dimensional relationships for the parts of the invention, to include variations in size, materials, shape, form, function and manner of operation, assembly and use, are deemed readily apparent and obvious to one skilled in the art, and all equivalent relationships to those illustrated in the drawings and described in the specification are intended to be encompassed by the present invention.

Therefore, the foregoing is considered as illustrative only of the principles of the invention. Further, since numerous modifications and changes will readily occur to those skilled in the art, it is not desired to limit the invention to the exact construction and operation shown and described, and accordingly, all suitable modifications and equivalents may be resorted to, falling within the scope of the invention.

We claim:

1. A sealed crustless sandwich, comprising:

a first bread layer having a first perimeter surface coplanar to a contact surface;

at least one filling of an edible food juxtaposed to said contact surface;

a second bread layer juxtaposed to said at least one filling opposite of said first bread layer, wherein said second bread layer includes a second perimeter surface similar to said first perimeter surface;

a crimped edge directly between said first perimeter surface and said second perimeter surface for sealing said at least one filling between said first bread layer and said second bread layer;

wherein a crust portion of said first bread layer and said second bread layer has been removed.

2. The sealed crustless sandwich of claim **1**, wherein said crimped edge includes a plurality of spaced apart depressions for increasing a bond of said crimped edge.

3. The sealed crustless sandwich of claim **2**, wherein said crimped edge is a finite distance from said at least one filling for increasing said bond.

4. The sealed crustless sandwich of claim **3**, wherein said at least one filling comprises:

a first filling;

a second filling;

a third filling; and

wherein said second filling is completely surrounded by said first filling and said third filling for preventing said second filling from engaging said first bread layer and said second bread layer.

5. The sealed crustless sandwich of claim **4**, wherein said first filling and third filling have sealing characteristics.

6. The sealed crustless sandwich of claim **5**, wherein:

said first filling is juxtaposed to said first bread layer;

said third filling is juxtaposed to said second bread layer; and

an outer edge of said first filling and said third filling are engaged to one another to form a reservoir for retaining said second filling in between.

7. The sealed crustless sandwich of claim **6**, wherein said first filling and said third filling are comprised of peanut butter; and said second filling is comprised of a jelly.

8. The sealed crustless sandwich of claim **7**, wherein said crimped edge is formed into a substantially circular shape.

9. A sealed crustless sandwich, comprising:

a first bread layer having a first perimeter surface, a first crust portion and a first contact surface;

a first filling juxtaposed to said first contact surface;

a second bread layer having a second perimeter surface, a second crust portion and a second contact surface;

a second filling juxtaposed to said second contact surface;

a third filling;

a crimped edge directly between said first and second perimeter surfaces for sealing said first, second, and third fillings between said first and second bread layers;

wherein said first and second crust portions have been removed and said third filling is encapsulated by said first and second fillings.

10. The sealed crustless sandwich of claim **9** wherein said first filling and said second filling have sealing characteristics.

* * * * *

Appendix C

Inventors and Inventorship

It is important, especially for scientists, engineers, and managers, to appreciate the unique nature of a patent and to understand the inventor's nature. A patent is not a technical paper; it is in fact quite different from such. And a technical person who labors long and hard to bring an invention to life is not necessarily an inventor. In this chapter, we will look at some inventors and study inventorship from social and legal points of view.

Who Are Inventors?

Patents come from inventions and inventions come from inventors. What sort of people are they? Do most inventors have characteristics in common? It has been alleged that inventors are dissatisfied with the technological status quo—they see it as needing improvement. According to Henry Petroski, in his book *The Evolution of Useful Things* (Vintage, 1994), they

> *appear to share the quality of being driven by the real or perceived failure of existing things or processes to work as well as they might. Faultfinding with the made world around them and disappointment with the inefficiency with which things are done appear to be common traits among inventors and engineers generally. They revel in problems—those they themselves identify in the everyday things they use, or those they work on for corporations, clients, and friends. Inventors are not satisfied with things as they are; inventors are constantly dreaming of how things might be better.*

So, then, are all inventors the same? Hardly. An important difference is to be found in the setting of the inventor.

The traditional, classic picture is the lone, laboring soul working in an isolated workshop. The inventor is portrayed as driven by a vision, forsaking the physical pleasures of life, unable to take respite from that which drives him or her.

The more collegial, more social setting is a research center laboratory. These inventors seem to differ from their traditional counterparts in a number of ways. First, their job appears to fill more nearly regular hours. Second, the scope of invention tends to be oriented more towards improvement of a technical product or process rather than devising something entirely new. The goals of these inventors are generally not set by the inventors themselves but by the business schema supporting and surrounding them. As such, the goals are narrower, sharper, and, perhaps, appear to some as less ambitious.

Certainly the products of the latter type of invention are generally less flashy than those of the former, and by themselves command little attention. One of these inventor types spoke for many of us when he said, "Most of what I have invented is embodied in hidden modules inside of large machines." This is the rhythm of the lion's share of invention activity today. Indeed, it is this latter setting of invention that, as early as 1904, inspired Daniel DeLeon, a socialist, to write "The poor lone inventor whom socialism would rob is a myth, and, like all myths, is adorned with the poetry of bygone ages. Like modern industry, modern invention is social. And, since the social characteristic of industry demands the social ownership of the means of production and distribution forming its basis, so does the social nature of invention demand the social ownership of its creations." ("Lo, the Poor Inventor!" in *The Daily People*, April 25, 1904)

Whatever the sentiment, the balance of invention has indeed shifted to the social context, and this is perhaps best marked by the year 1932. It was then that corporate patent assignees first outnumbered individual assignees.

Failing Your Way to Success

Asking inventors to summarize the invention process or to explain why they are prolific inventors is, more often than not, destined to result in an answer that does not seem deep, mysterious, or exceptionally helpful. What the answer does often convey is a hint of the creative force that lives through the inventor and will not be suppressed.

Consider another of Edison's statements regarding invention: "Genius? Nothing! Sticking to it is the genius! Any other bright-minded fellow can accomplish just as much if he will stick like hell and remember nothing that's any good works by itself. You've got to make the damn thing work!" And, "I failed my way to success."

Highly successful inventors are not alone in this regard. In 1864, then Commander Ulysses S. Grant was asked about the art of war. He replied, simply, "Find out where your enemy is, get at him as soon as you can and strike him as hard as you can, and keep moving on." Perhaps the common denominator of great producers, the essence of what makes them special, is that they are possessed of (or are possessed by) an almost driven, indefatigable striving, and if you work with or manage inventors, it will do you good to recognize that for all of their ability, for all of their value, many of them have fragile egos and, strangely, feelings of inadequacy and significant need for continual justification. An effective manager will take the time to deal with these individual issues and be sure that the process is properly psychosocially "tuned."

One prominent quirk of many inventors is a premature fear of non-originality. Often solutions seem obvious to inventors, while those same solutions are actually and legally nonobvious to persons of ordinary skill in the art. For this reason, inventors sometimes tend to take a little information, decide that a problem has already been solved and that their efforts are not original, and abandon efforts that might realize exceptional value for the company.

If you have read this entire book, you should now have some skill in examining patents. Try the following experiment with an inventor who is not particularly schooled in interpreting patents: Choose a field with which the inventor is familiar. Find an issued patent in that field. Without first showing the issued patent to the inventor, ask the inventor to attempt invention along the lines of the issued patent. Then ask the inventor to read the first few claims of the issued patent and tell you what he or she thinks. The inventor will likely believe that there is no room for invention and will read the words of the claims onto what he or she had invented. However, the patent probably doesn't actually completely negate the patentability of the inventor's invention.

Success Has Many Fathers

Another quirk is fear of diminishment. A patent is not really a work of art. It is not an official recognition of individuality. It is, of course, an official recognition and deed to something new, useful, and nonobvious. Its purpose is not to praise or distinguish the inventor, not even as someone of above ordinary skill in the art—as the inventor's skill is a nondeterminative factor in the novelty analysis. But the patent does insist on publishing the inventor's name and linking it with the new, useful, and nonobvious *something*. It is here that an important aspect of human nature—the pride of authorship—is revealed.

It is characteristic of many inventors to want to be perceived, or perhaps to perceive themselves, as a key creative force; because of this, many inventors pointedly disdain encouraging other inventors to join their patent application. But by law, if someone significantly contributes only to a single claim of an application, that person must be listed as an inventor on that application, a co-equal position to an inventor who might have done all of the rest. For many inventors, having to include another, or others, is to admit that his or her efforts were incomplete, and this is anathema to many of them.

One of the most difficult challenges for highly educated inventors is to understand that a patent and a peer-reviewed paper are very different despite their similarities. The patent lists the inventors. The paper lists the authors. Exclusive of naked plagiarism, inclusion or exclusion of someone from the list of authors to a paper is not a matter of law, but a matter of courtesy and culture. In a patent, the list of inventors must be scrupulously accurate, neither including noninventors nor excluding actual inventors. In a paper, the order of the authors is often taken to be a measure of relative prestige or weight of contribution. In a patent, the order of the listing of the inventors is meaningless. In a paper, the subject matter has to be explained or set in the particular art's context. In a patent, explanation, beyond instruction of how to make and use the invention, is not required.

The real importance of the patent is not, of course, to the inventors but to the assignee or company. It must fall to management to be sure that the company's patenting activity results in the greatest benefit to the company, and this means dealing positively and persuasively with inventor egos, fantasies, and fears.

Recognition and Compensation

Are inventors the same as most of our colleagues who chase money and power as the metric of their success, of their worth? Certainly inventors appreciate a certain special remuneration for their labors, and in order for the assignee to *clear the title* to their inventions, so to speak, they are often given a consideration for their work. But the consideration is often little more than a legal transaction to the typical inventor. Legally, of course, the consideration is, as Black's *Legal Dictionary* puts it, "the motivation to a contract," and some believe that the inventor's salary itself sufficiently meets such a test. But the inventor, by and large, is removed from the argument. The inventor wants to invent; it is a key part of the inventor's nature and to deny it is to frustrate and lose intellectual capital.

This is not to say that the inventor doesn't enjoy or seek a type of compensation. It's just not generally money. It might be, and often is, recognition, and to bestow this recognition and thereby stroke the inventor, companies have adopted a variety of mechanisms. Some companies have erected a photo display of prolific inventors that feature various levels according to the number of issued patents. Another approach is to render a prodigious inventor's badge slightly but noticeably different. And then, of course, there are celebration dinners and outings—a challenge for human resources, because the reward paradigm must be tailored and not stamped out.

The 38 respondents of a survey reported by Japan Intellectual Property Association (JIPA) Forum in 2004 (by J. Jeffrey Hawley) reported the following:

- 84 percent were awarded no payment for an invention disclosure.
- 5 percent were awarded $1–$100.
- 11 percent were awarded $101–$200.

Of the responding companies,

- 39 percent were awarded no payment for filing an application.
- 21 percent were awarded $1–$500.
- 29 percent were awarded $501–$1500.
- 11 percent were awarded $1501–$3000.

The same group also reported on payments for patent issuance:

- 63 percent were awarded no payment.
- 26 percent were awarded $1–$500.
- 11 percent were awarded $501–$2000.

A typical employment agreement reported in the JIPA report speaks directly to ownership of the employee's inventions:

> *I understand that the Company requires its employees to assign to it all right, title and interest in all inventions, discoveries, improvements and copyrightable subject matter within or arising out of any field of employment with the Company. Therefore, in consideration of my employment by the Company and of the wages and other benefits to be received by me in connection with such employment, it is understood and agreed as follows: I hereby sell, assign and transfer to the Company all of my right, title and interest in all inventions.... which during my employment and within (X) years following my termination, are made or conceived by me alone or with others... I will fully disclose to the Company as promptly as available all information known or possessed by me concerning the inventions...referred to above, and upon request by the Company and without further remuneration in any form to me by the Company will execute all documents necessary to perfect rights to the inventions.*

The Meaning of "Inventorship"

Inventorship is a critical item to the USPTO. If inventorship is not correctly recorded, if someone who was an inventor is not listed, or, conversely, if someone is listed as an inventor who was not an inventor, then the patent's validity might be assailed. Inventorship starts with mental conception and becomes patentable after the concept is taught so that someone of ordinary skill in the art may build and practice the invention.

The *Manual of Patent Examining Procedure* (MPEP) §2138.04 states the following:

> *Conception has been defined as "the complete performance of the mental part of the inventive act" and it is "the formation in the mind of the inventor of a definite and permanent idea of the complete and operative invention as it is thereafter to be applied in practice."*
>
> *The* inventor *must form a definite and permanent idea of the complete and operable invention to establish conception.*

In the court case Pfaff v Wells Elecs., Inc., 525 U.S. 55, 55 (1998), the court wrote that "The primary meaning of 'invention' in the Patent Act unquestionably refers to the inventor's conception rather than to a physical embodiment of that idea. The statute contains no express 'reduction to practice' requirement…and it is well settled that an invention may be patented before it is reduced to practice."

An inventor eligible for a patent is someone who has contributed significantly to at least one claim of the invention. This is an important concept and is sometimes divisive, because many people might work on an invention but only a subset of those people will be actual inventors. For example, inventors A and B might have become aware of a newly reported electronic phenomenon, perhaps the behavior of a material under certain environmental or electrical conditions, and they conceive of a new electronic device exploiting such. Their boss, C, might suggest that they construct a prototype to see if a new electronic device exploiting the phenomenon can be built, and the boss then provides funding for the research and development. A and B are successful in this endeavor. D joins them on the project in computer simulation of the device's behavior. E joins in helping to construct the prototype. Finally, A and B might not understand exactly why the device works, and F joins the group and works out a theory to explain the new effect and why the device works. For purposes of a patent, only A and B likely qualify as inventors. D and E are not likely inventors, because according to MPEP 2137.01, "it is not essential for the inventor to be personally involved in carrying out process steps, where implementation of those steps does not require the exercise of inventive skill."

The boss, C, is not likely to be an inventor as, according to MPEP 2137.01, "one who suggests an idea of a result to be accomplished, rather than the means of accomplishing it, is not a coinventor." F also is probably not an inventor, because it is not necessary for the inventors to understand exactly why their invention works; in fact, they might even have an incorrect theory of operation. What is essential is for the inventors to describe and enable their invention, to teach someone of ordinary skill in the art how to realize the invention and how to practice it.

Additionally, A and B need not have contributed equal amounts to the invention for both to be inventors. According to the ruling in Ethicon, Inc. v U.S. Surgical Corp., 135 F.3d 1456 (Fed. Cir. 1998),

> *When an invention is made by two or more persons jointly, they shall apply for patent jointly and each make the required oath, except as otherwise provided in this title. Inventors may apply for a patent jointly even though (1) they did not physically work together or at the same time, (2) each did not make the same type or amount of contribution, or (3) each did not make a contribution to the subject matter of every claim of the patent.*

A very good summarization of many key points of the law surrounding inventorship, and joint inventorship in particular, is found in the Ethicon case under Section II "Co-Inventorship" (legal citations removed and bulleting added):

- *A patented invention may be the work of two or more joint inventors. Because "[c]onception is the touchstone of inventorship," each joint inventor must generally contribute to the conception of the invention. "Conception is the 'formation in the mind of the inventor, of a definite and permanent idea of the complete and operative invention, as it is hereafter to be applied in practice.'" An idea is sufficiently "definite and permanent" when "only ordinary skill would be necessary to reduce the invention to practice, without extensive research or experimentation."*
- *The conceived invention must include every feature of the subject matter claimed in the patent. Nevertheless, for the conception of a joint invention, each of the joint inventors need not "make the same type or amount of contribution" to the invention. Rather, each needs to perform only a part of the task which produces the invention. On the other hand, one does not qualify as a joint inventor by merely assisting the actual inventor after conception of the claimed invention. "An inventor 'may use the services, ideas and aid of others in the process of perfecting his invention without losing his right to a patent.'" One who simply provides the inventor with well- known principles or explains the state*

of the art without ever having "a firm and definite idea" of the claimed combination as a whole does not qualify as a joint inventor. Moreover, depending on the scope of a patent's claims, one of ordinary skill in the art who simply reduced the inventor's idea to practice is not necessarily a joint inventor, even if the specification discloses that embodiment to satisfy the best mode requirement.

- *Furthermore, a co-inventor need not make a contribution to every claim of a patent. A contribution to one claim is enough. Thus, the critical question for joint conception is who conceived, as that term is used in the patent law, the subject matter of the claims at issue.*

A widely held belief is that anyone who makes a contribution to a patent claim qualifies as an inventor. This rule is almost always true, but exceptions occur occasionally, and these can be highly significant in the course of business affairs and their litigation. Such was the case in Nartron Corp. v Schukra U.S., Inc. (Federal Circuit 2008-*1363*, March 5, 2009).

The question before the court hinged on whether or not a Mr. Benson, a Schukra employee, was a co-inventor of a patent assigned to Nartron. The problem arose because Nartron was suing a Schukra supplier for infringement, and a district court had granted summary judgment of dismissal of Nartron's suit because neither Schukra nor Benson had been joined, or consented to be joined, to Nartron's suit as would be required for a co-inventor. It was therefore in Nartron's best interest to have Benson declared not a co-inventor so that the case could be remanded to the district court. The case centered on Claim 11, a dependent claim in a chain of dependent claims pending from Claim 1:

1. A seat control module for introducing massage to a seat control with an adjustable lumbar support, and control actuators, the control module comprising:

a modular housing including in-line connectors for coupling said module to a seat control harness connector;

an intercept interface for receiving inputs from said control actuators;

a driver for repeatedly adjusting said lumbar support position through a predetermined range of movement in response to one of said control actuators; and

a transparency simulator for maintaining full function of said seat control and removing indications of repeatedly adjusting said lumbar support position....

5. The invention as defined in claim 1 wherein said transparency simulator comprises time-based response to manipulation of said control actuators.

6. The invention as defined in claim 5 wherein said transparency simulator generates a first output in response to a manipulation of a said control actuator for a period of time less than a first predetermined period, and generating a second output in response to a manipulation of said control actuator for a period equal to or greater than said first predetermined time period....

11. The invention as defined in claim 6 wherein said lumbar support adjustor includes an extender.

The district court held the extender element in claim 11 was Mr. Benson's conception. Schukra's supplier maintained that Mr. Benson's contribution was not insignificant and that "the subject matter of dependent claims must be invented by named inventors" and that "an aspect of a claim cannot be insignificant simply because it appears in a dependent claim."

Schukra's supplier conceded that "a lumbar support adjustor with an extender existed in automobile seats in the prior art" and the court concluded that the "contribution of supplying the extender to the patented invention was the exercise of ordinary skill in the art." In the end, the court decided that the alleged co-inventor provided only an insignificant contribution to the invention of claim 11 of the patent by contributing an extender and reversed and remanded. The court also made a further point:

A dependent claim adding one claim limitation to a parent claim is still a claim to the invention of the parent claim, albeit with the added feature; it is not a claim to the added feature alone. Even if Benson did suggest the addition of the prior art extender to what Nartron had invented, the invention of claim 11 was not the extender, but included all of the features of claims 1, 5, and 6, from which it depends.

Inventing as a Passion

For many, inventing is simply a passion, something they are driven to do by the force of things that interest them. Here are a few portraits drawn from history of people who were passionate about finding new and better ways.

Abraham Lincoln

In the early 1800s, in Sangamon County, Illinois, waterways were essential to life and commerce, so it is not surprising that a creative personality would be attuned to innovating methods and apparatuses for river vessels. The 16th U.S. president was such a person. Lincoln's application for patent of a device for "Buoying Vessels Over Shoals," was filed on March 10, 1849, and granted as patent 6469 on May 22 of that year to the only U.S. president to hold a patent. His patent has one claim:

> *What I claim as my invention and desire to secure by letters patent, is the combination of expansible buoyant chambers placed at the sides of a vessel, with the main shaft or shafts C, by means of the sliding spars, or shafts D, which pass down through the buoyant chambers and are made fast to their bottoms, and the series of ropes and pullies [sic], or their equivalents, in such a manner that by turning the main shaft or shafts in one direction, the buoyant chambers will be forced downwards into the water and at the same time expanded and filled with air for buoying up the vessel by the displacement of water; and by turning the shaft in an opposite direction, the buoyant chambers will be contracted into a small space and secured against injury.*

Lincoln was evidently quite interested in mechanical items and very appreciative of the patent laws. That he had a passion for invention is seen not only by his inventive quest but also in his writings. Two of his quotations are often cited in support of this claim: "Man is not the only animal who labors; but he is the only one who improves his workmanship." And "The patent system added the fuel of interest to the fire of genius."

Albert Einstein and Leo Szilard

These mental giants were key contributors to the theory of mass-energy equivalence, the concept of chain reaction, and ultimately the development of the atomic bomb. But they also jointly worked on inventions for refrigeration, and their motivation for some of this work was tied to a sympathy and passion for human safety.

From 1924 to 1934, Szilard applied to the German patent office for 29 patents both individually and jointly with Einstein. Most of the joint applications dealt with home refrigeration after a sad newspaper story caught the inventors' attention

A Berlin newspaper reported that an entire family had been found asphyxiated in their apartment as a result of their inhalation of noxious fumes of a chemical used as the refrigerant in their primitive refrigerator that had escaped in the night through a leaky pump valve. Moved by the story, the two physicists devised a method of pumping metallicized refrigerant by electromagnetism, a method that required no moving parts (and therefore no valve seals that might leak), except the refrigerant itself.

Kelly Fitzpatrick

"It's the only real thing I'd wanted to do my whole life." That's what Kelly Fitzpatrick said about hairdressing. As a teenager she injured a hand, and that kept her from her desired vocation. She worked instead in real estate, but she kept returning to hair shows. While watching a TV show for aspiring inventors, she dedicated her inventive spirit to solving one of hairdressing's biggest concerns, hair volumizing.

It took a lot of trial and a lot of time, but Fitzpatrick finally perfected her product, called the Bumpit. She filed a number of patent applications. U.S. Patent 2007888205 is one of the earlier ones, with the application publishing in early 2009. The background does an admirable job of motivating the invention's utility:

> *One desirable hair treatment or style is to manipulate the hair so as to appear to increase the volume of a person's hair. In particular, many women desire to have hair that is full of volume so as to achieve a look that is commonly referred to as the "big hair" appearance. Generally, achieving this look requires the hair to be teased or bumped…and then finished to hold in place. Most women have much difficulty in obtaining this look on their own and, as a result, must go to a hair stylist to obtain the desired volumizing of their hair.*

And the summary answers the background:

> *The present invention discloses a hair volumizing device which is worn in the hair and useful for providing the appearance of increased volume in the user's hair. The hair volumizing device of the present invention is an easy to use device that can be placed by the user in her own hair and then worn under the hair to provide the appearance of increased volume without the device being seen by others. The present hair volumizing device is adapted to engage a portion of the user's hair just above the scalp and raise the hair up to give the user the appearance of increased hair volume without the need to see a*

hairstylist or tease/bump the hair themselves. The preferred hair volumizing device of the present invention is made out of components and materials that can provide a relatively inexpensive to manufacture device.

Recap

Establishment of correct inventorship is extremely important, and not getting it right can imperil the patent's validity. Inventorship belongs to those who conceived the invention. The patent claims are the key. If someone has made a significant contribution to any claim, that person is one of the inventors. If a person has not made a significant contribution, then he or she should not be claimed as one of the inventors.

Discussions and Reflections

- George de Mestral lived from 1907 to 1990. An electrical engineer, he invented Velcro ("vel" from *velour*, French for velvet; and "cro" from *crochet*, French for hooks). Research de Mestral's life and learn how his passion for invention of this remarkable material started by observing something in nature. Then see if you believe what he supposedly told some executives: "If any of your employees ask for a two-week holiday to go hunting, say yes." Think about other inventions that might be inspired by nature. How about viruses and other malware and immunizations to deal with these?
- If your company is planning to acquire patent rights, it is important that their strength be assessed under test, because patent value can significantly depend on correct and complete representation of inventorship. How do your colleagues and employees conduct this part of the M&A procedure?
- Review the inventor's compensation mechanism in your company.
  - How do the inventors view it?
  - Is there a way to measure the compensation mechanism's effectiveness?
  - Is the invention consideration policy uniform throughout the company? Should it be?

Index

M

N

O

P

Q

R

S

T

U

V

W

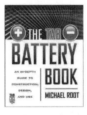

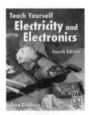

Masterpieces *of the* J. Paul Getty Museum

ILLUMINATED MANUSCRIPTS

Masterpieces *of the* J. Paul Getty Museum

ILLUMINATED MANUSCRIPTS

Los Angeles
THE J. PAUL GETTY MUSEUM

Frontispiece:
Master of the Dresden Prayer Book,
The Temperate and the Intemperate
[detail] (See no. 41)

At the J. Paul Getty Museum:

Christopher Hudson, *Publisher*
Mark Greenberg, *Managing Editor*
Mollie Holtman, *Editor*
Suzanne Watson Petralli, *Production Coordinator*
Charles Passela, *Photographer*

Text prepared by Thomas Kren, Elizabeth C. Teviotdale,
Adam S. Cohen, and Kurtis Barstow

Designed and produced by Thames and Hudson
and copublished with the J. Paul Getty Museum

1200 Getty Center Drive
Suite 1000
Los Angeles, California 90049-1687

Library of Congress Card Number 97-070932

ISBN 0-89236-446-7

Color reproductions by CLG Fotolito, Verona, Italy

Printed and bound in Singapore by C.S. Graphics

DIRECTOR'S FOREWORD

The collection of illuminated manuscripts covered by this book, like so much else about the Getty Museum, is a new creation, having been formed only in the past fifteen years. When J. Paul Getty's will was opened in 1976, it was discovered that he had made a seven hundred million dollar bequest to his museum. A small institution with a narrow, uneven collection was suddenly in a position to expand in any direction its trustees chose. During the six years in which lawsuits prevented them from using the legacy, new possibilities were explored by the Getty Trust for services to scholarship, conservation, and arts education, as well as for building up a much more important museum.

When I decided to come to the Getty in 1983, one idea for expanding the Museum's collection was already in the works: the acquisition *en bloc* of the Ludwig Collection of illuminated manuscripts. Getty's own interests as a collector had been confined to antiquities, decorative arts, and paintings, and the Museum had not strayed outside those boundaries. These illuminated manuscripts offered a chance not only to annex the Middle Ages and early Renaissance, but also to show the public a vast array of brilliantly preserved pictures that would never be rivaled by later purchases of panel paintings.

At the urging of Thomas Kren, then Associate Curator of Paintings, the acquisition was made. Soon we created the Department of Manuscripts with Dr. Kren as its first curator; a staff was recruited, a study room fitted out, and an ambitious program of activities launched. These have included regular exhibitions, catalogues of the permanent collection, scholarly studies, and exhibition catalogues—taken together, a large achievement for such a young department. At the same time, new collections of drawings, sculpture, and photographs were also formed; these have tremendously enriched our visitors' experience in the past dozen years.

Added to the Ludwig Collection have been purchases of manuscripts and cuttings, among them many of our greatest works. These will be published in a catalogue by Thomas Kren scheduled to appear in the near future.

To the writers of this book—Thomas Kren, assisted by Elizabeth C. Teviotdale, Adam S. Cohen, and Kurtis Barstow, all of the Department of Manuscripts—I am very grateful.

Works of art are always distorted by reproductions in books, which shrink them into patches of printer's ink. The distortion is least in the case of manuscript illumination. We hope that turning the pages of this book will offer the reader at least some of the joys of close contact with the originals and will be an incentive for a visit to the new Getty Museum, where every day many of our finest manuscripts can be seen.

JOHN WALSH
Director

INTRODUCTION

The Getty Museum's endeavor to create a collection representative of the history of European manuscript illumination is atypical. Despite their important place in the history of European art, illuminated manuscripts have not found their way into art museums as most other portable artistic media have. Relatively few of the art museums created during the nineteenth and twentieth centuries in Europe and America have actively collected illuminated manuscripts. This is in part because these lavish books have generally passed from private libraries into public ones. Many of the great imperial, royal, ducal, and even papal manuscript collections became components of national and state libraries. This pattern has generally held into the twentieth century. The American collector J. Pierpont Morgan (1837–1913) gave generously of his splendid holdings in medieval art to museums, principally to the Metropolitan Museum of Art in New York. But his extraordinary collection of medieval and Renaissance illuminated manuscripts was not included in that gift, instead becoming part of the private library that carries his name. Therein resides the finest collection of medieval painting in America. The greatest museum repository of illuminated manuscripts, the British Museum, London, handed over its illuminated books to the newly formed British Library only twenty-five years ago.

Conversely, the Metropolitan Museum of Art, with its resplendent collection of medieval art, has acquired only a handful of illuminated codices (albeit magnificent and important ones) along with a select group of leaves. It has refrained from collecting illumination actively or systematically. Only two people envisioned collecting illuminated manuscripts within the context of encyclopedic art collections: Henry Walters, whose collections formed the basis for the Walters Art Gallery in Baltimore, and William Milliken, Director and Curator of Medieval Art at the Cleveland Museum of Art, with the Walters concentrating on books and Cleveland on cuttings.

The Getty Museum's collection of illuminated manuscripts owes something to both the Walters and Cleveland models. Although it became apparent early, due both to issues of cost and availability, that the modern ideal of encyclopedic collections would not be feasible at the Getty, the trustees proposed expanding the Museum's collections beyond the three areas to which J. Paul Getty (1892–1976) had limited himself. Medieval art was one of the targeted fields.

The collection was begun with the purchase of 144 illuminated manuscripts assembled by Peter and Irene Ludwig of Aachen, Germany, in 1983. The finest collection of illuminated manuscripts formed in the second half of the twentieth century, the Ludwigs' holdings were among the very few private collections of the material still intact. Their collection was selected with the advice of book dealer

Hans P. Kraus to provide a historical survey of the illuminated manuscript, representing a broad time frame and range of schools along with great variety in the types of books. This purchase of the Ludwig manuscripts not only added a number of masterpieces of medieval and Renaissance art to the Museum's collection but also complemented that of European paintings, extending the coverage of the history of painting back to the ninth century. Since then the department has added to these holdings selectively, filling gaps and building on strengths where possible.

The following pages display illuminations from the Museum's finest manuscripts, including a number of fragmentary ones. The selections are arranged roughly chronologically in a survey that reflects broadly the strengths of the collection. The book commences with a Gospel lectionary from the late tenth century, produced in one of the great monastic scriptoria of the Ottonian era. It concludes with an unusual Model Book of Calligraphy with scripts by Georg Bocskay, imperial court secretary to the Hapsburg emperor Ferdinand I (r. 1556–64), and illuminations by Joris Hoefnagel, a court artist for the emperor Rudolf II (r. 1576–1612). The six hundred years that separate these two books witnessed tremendous social and cultural changes, including the transition from monastic to lay workshops of book production, an explosion in book collecting resulting in the formation of the great court libraries, the growth of aristocratic patronage, and the emergence of bourgeois patronage.

By way of illustration, the deluxe Gospel book enjoyed its widest appeal in Western Europe only until the twelfth century, but reigned for a much longer period in the eastern Mediterranean and Christian Near East. Although it was unimportant before the thirteenth century, the book of hours found an audience of lay people whose scope would have been unimaginable in the earlier era, and so it became the most popular book of the later Middle Ages and a primary vehicle for illuminators' artistic innovations in Western Europe. Other consequential changes over the course of the Middle Ages and Renaissance include the rise of vernacular literature and translations along with an expansion in the types of works that were deemed appropriate to illuminate; a new level of self-consciousness on the part of artists; and the introduction of printing from movable type that would result eventually in the hegemony of the printed book.

The commentaries herein seek to highlight some of these changes. The reader will find, for instance, that in the later period we not only have a broader range of titles but generally know more about both the artists and the patrons. Whenever possible, the authors of the entries have provided a wider artistic or historical context for the manuscript; inevitably, where the collection is stronger, more connections are drawn among the examples.

The selection of objects in this book surveys the cream of the collection, but it is not a proper historical survey. Rather, it reflects the imbalances within the Museum's holdings. The Ludwig manuscripts enjoy particular riches among German illumination of the eleventh to the fifteenth centuries and in late medieval Flemish manuscripts. The Getty has added examples to these areas over the years while actively developing the collection of later medieval French illumination, an area in which the Ludwigs acquired only a few manuscripts. At the same time certain schools from the beginning of the tradition (Early Christian to Carolingian) remain thinly represented or absent. But part of the challenge of collecting as well as the pleasure is the discovery of the rare and unexpected. The goal of assembling a representative and balanced history of the art of manuscript illumination is doubtless unattainable. Nevertheless, it focuses our efforts to improve the overall quality and character of the Getty's small collection. Perhaps time, perseverance, and fortune will permit us to fill some of the lacunae.

THOMAS KREN
Curator of Manuscripts

NOTE TO THE READER

The following types of illuminated manuscripts appear in this book:

untitled books
titled books
groups of leaves from an identifiable manuscript
miniatures and historiated initials from an identifiable manuscript
miniatures from an unidentifiable manuscript
miniatures that may or may not come from a manuscript.

In the case of untitled books, generic titles are used (e.g., book of hours).

Titled books are cited by author and title (in the original language) or by title alone.

In the plate captions, the artist's name is given when known. The medium for all of the painted decoration in the manuscripts featured in this book is tempera colors, sometimes used together with gold leaf, silver leaf, or gold paint. The support is customarily parchment, although the sloth in Ms. 20 (no. 53) is painted on paper.

We cite the Bible according to the Latin Vulgate version.

1 Gospel Lectionary
Saint Gall or Reichenau, late tenth century

212 leaves, 27.7 x 19.1 cm ($10^{15}/_{16}$ x $7^{9}/_{16}$ in.)
Ms. 16; 85.MD.317

Plate: Decorated Initial *C*, fol. 2

The reconstitution of Charlemagne's Holy Roman Empire under Otto I in 962 ushered in a new era of luxury art production. The imperial dynasty of Saxon kings in Germany dominated the European political landscape from the mid-tenth into the eleventh century. Over the course of roughly one hundred years some of the most sumptuous illuminated manuscripts of the Middle Ages were produced in the Ottonian realm.

The lectionary is a collection of the Gospel selections to be read during the mass. As in many such liturgical manuscripts, the most important feast days are marked by pages filled with large and elaborate decorated initials (called incipit pages) to introduce the readings. This ornate *C* begins the Latin of Matthew 1:18, "When Mary his mother was espoused to Joseph . . ." (*Cum esset* . . .), the passage read on Christmas eve.

The style of the foliate initial indicates that this manuscript was created in either Saint Gall or Reichenau (both near the modern Swiss-German border). The monasteries there were among the first centers for the production of Ottonian manuscripts, and the rich use of gold and purple in this and other works reveals the wealth possessed by such religious foundations. Reichenau in fact was noted for the creation of opulent manuscripts associated with the imperial house, while Saint Gall had a long tradition of scholarship and art production reaching back to the age of Charlemagne and beyond.

Although initials had been given prominence in earlier medieval manuscript painting, Ottonian pages display an unprecedented and remarkable formal harmony. In this characteristic example, the rectangular frame creates a clearly defined space for the initial and serves as an anchor for the golden vines that intertwine with the letter *C*. Within the frame, the brilliant gold is subtly accented by pale patches of blue and lavender with spots of orange and dark blue, all of which is contrasted with the creamy tone of the parchment. Greatly admired for their beauty and rarity, few Ottonian manuscripts are to be found in American collections, and in this respect the Getty Museum's group of four books is exceptional (nos. 1, 3–5). ASC

CVM
ESSET

2 Two Leaves from a Gospel Book
Canterbury (?), circa 1000

31.3 x 20.2 cm (12⁵⁄₁₆ x 8 in.)
Ms. 9; 85.MS.79

Plate: *The Miracle of the Stater,* leaf 2

The Museum's two Anglo-Saxon leaves come from an illuminated Gospel book (a book containing the accounts of Christ's life written by Saints Matthew, Mark, Luke, and John). They include three full-page miniatures, all illustrating episodes from the miracles and ministry of Christ that are rarely found in early medieval art.

The story of the miracle of the stater is briefly outlined in the Gospel of Saint Matthew (17:26): at Capharnaum, Jesus instructed Saint Peter to go to the sea and cast his hook so that he might find in the mouth of the first fish he caught a coin (or stater) with which to pay a tax. The Anglo-Saxon illuminator has presented the story in two scenes that imply three moments in the story (leaf 2). The scene above seems to show both Jesus instructing Peter—conveyed in Jesus' gesture—and Peter returning with the stater, while the scene below shows Peter catching the fish. Thus the illuminator cleverly plays on our expectation that the scenes should be read from top to bottom and instead presents a narrative that moves from the upper scene to the lower scene and back up.

The visual interest of the miniature is enhanced by the illuminator's lively drawing style, one favored in late Anglo-Saxon manuscript painting. This technique has its distant roots in the impressionism of ancient Roman painting, but Anglo-Saxon artists exploited its expressive potential more than their ancient predecessors had. The agitated drawing is especially effective in the depiction of the surface of the water and the wriggling fish on Peter's hook. ECT

3 Sacramentary
Fleury, first quarter of the eleventh century

10 leaves, 23.2 x 17.8 cm
(9⅛ x 7 1/16 in.)
Ms. Ludwig V 1; 83.MF.76

Plate: Attributed to Nivardus of Milan, Decorated Initial *D* with Clambering Figures, fol. 9

A sacramentary is a book containing the prayers recited by the celebrating priest at mass—the Christian rite in which bread and wine are consecrated and shared. Serving as a part of the adornment of the altar during mass, sacramentaries were often illuminated in the early Middle Ages, especially if they were made for presentation to powerful political or ecclesiastical officials.

This sacramentary, only a fragment of which is preserved, may have been made for Robert the Pious, King of France (r. 996–1031), perhaps at the behest of the Bishop of Beauvais, who crowned Robert in 1017. The writing and illumination have been attributed to Nivardus, an Italian artist who worked at the Benedictine monastery of Saint-Benoît-sur-Loire at Fleury in France. Nivardus's gold and silver initials were inspired by those of illuminators active at the monasteries of Saint Gall and Reichenau (see no. 1), but his initials are distinct in the abundance of the knot work. This exuberance of decoration sometimes obscures the shapes of the letters.

The initial *D* that introduces the prayers for Easter (fol. 9) is framed by a pair of columns surmounted by vines that complement the form of the initial. The decoration of the page is further enhanced by the inclusion of a pair of clambering figures, their poses and the colors of their clothing contributing to the lively and masterfully harmonized design of the page as a whole. ECT

S̃QVIHODIERNA DIE

4 Benedictional
Regensburg, circa 1030–1040

117 leaves, 23.2 x 16 cm
(9 1/8 x 6 5/16 in.)
Ms. Ludwig VII 1; 83.MI.90

Plate: *The Adoration of the Magi,* fol. 25v

Regensburg, the capital of Bavaria in the Middle Ages, was one of the most important political, religious, and cultural centers in all of Europe. The luxury manuscripts produced under the patronage of the Ottonian emperor Henry II (r. 1014–1024) attest to Regensburg's prosperity at this time, and for the rest of the century the city would remain the focal point of a flourishing culture that extended throughout the region.

This benedictional, a book containing the blessings recited by the bishop at mass, was made for Engilmar, who is depicted celebrating mass on the original opening page of the manuscript. Engilmar's career reflects the wide-ranging ties made possible by the network of Benedictine monasticism. First a monk in the monastery at Niederaltaich (Bavaria) and later the Bishop of Parenzo (modern Poreč in northwestern Slovenia across the gulf from Venice), Engilmar was an honored guest at Saint Emmeram, Regensburg's chief monastery. Stylistic comparisons to other manuscripts indicate that the bishop turned most likely to Saint Emmeram for the production of his benedictional, sometime between 1030 and 1040.

The Adoration of the Magi is one of seven full-page narrative scenes from the Life of Christ in the book and introduces the feast of Epiphany on January 6. The subject was one of the most popular in medieval art, and the benedictional's picture relies on earlier Ottonian art from Reichenau for its composition. The figures here loom large in relationship to the framing architecture, and they are highlighted by the gleaming gold background that reinforces the miraculous aspect of the event. The monumentality of the enthroned Virgin Mary is particularly striking as she and Jesus respond dramatically to the adoring kings. Such demonstrative hand movements are a quintessential trait of Ottonian art, in which the language of gesture found some of its most lyrical visual expression.

ASC

Per omnia scla sclorum. Amen.
Dns vobiscum. Et cum spu tuo.
Sursum corda. Habemus ad dnm.
Gratias agamus dno deo nro.
Dignum et iustum est.

5 Sacramentary
Mainz or Fulda, second quarter of the eleventh century

179 leaves, 26.6 x 19.1 cm
(10½ x 7 7/16 in.)
Ms. Ludwig V 2; 83.MF.77

Plates: *Pentecost* and Incipit Page, fols. 20v–21

See pages 18–19

Archbishop Bardo of Mainz (in modern Germany) probably gave this richly illuminated sacramentary, together with relics of Saint Alban (d. 406), to the Cathedral of Saint Alban at Namur (in modern Belgium) at the time of its foundation in 1046. The book, whose covers are embellished with metalwork and enamels, would have been kept in the cathedral's treasury and placed on the altar for use at mass only on important feast days.

The most remarkable artistic feature of this sacramentary is the series of six full-page miniatures of key events in New Testament history that precedes the main text. Such prefatory cycles are rare in early medieval liturgical manuscripts. The final miniature of the series shows the descent of the Holy Spirit on the apostles at Pentecost (fol. 20v). The miniature is a literal representation of the event as it is described in the Bible (Acts 2:1–4). The apostles are sitting in a house as "parted tongues of fire alight on each one of them." Although the inclusion of the roof sets the scene in a house, the gold background imparts an otherworldly character, emphasizing that the apostles "were all filled with the Holy Spirit."

The Pentecost miniature harmonizes with the text page opposite (fol. 21) through the shared colors of the large frames decorated with foliate motifs. The text, the opening of one of the prayers of the mass, is written in gold on a purple and green background. This treatment deliberately imitates the appearance of the most sumptuous manuscripts of the Roman imperial period, in which the texts were written in precious metals on purple-dyed parchment. ECT

6 Gospel Book
Helmarshausen, circa 1120–1140

168 leaves, 22.8 x 16.4 cm
(9 x 6½ in.)
Ms. Ludwig II 3; 83.MB.67

Plates: *Saint Matthew* and Incipit Page, fols. 9v–10

See pages 22–23

The Gospels, the accounts of Christ's life attributed to Saints Matthew, Mark, Luke, and John, lie at the center of Christian teaching. From the seventh to the twelfth century the most important and beautiful illuminated manuscripts produced in Western Europe were Gospel books. This twelfth-century example, one of the finest manuscripts in the Getty Museum's collection, was produced at the Benedictine Abbey at Helmarshausen in northern Germany.

Each Gospel is prefaced by a portrait of the author, a pictorial tradition that originated in antiquity. Here we see Saint Matthew writing the opening lines of the text: *Liber generationis jesu christi filii David filii habrah[am]* (The book of the generations of Jesus Christ son of David son of Abraham). The inscription above Matthew's head reads "The beginning of the Holy Gospel according to Matthew." The writer holds a quill pen and a knife to sharpen it. Two ink-filled horns are set into the lectern.

The large areas of rich color and the pattern of folds of the bulky robe are particularly characteristic of Romanesque art. The folds are simplified into geometric shapes and frequently "nested," that is, set neatly within one another. Despite this stylization, Matthew is a robust, full-bodied figure. As is usually the case with illuminated manuscripts from the Middle Ages, we do not know the artist who painted these pages. The eminent metalsmith Roger of Helmarshausen, who was active in Lower Saxony at the beginning of the twelfth century, designed figures in a strikingly similar manner.

The incipit, or opening lines of a text, often received artistic attention equal to that given the miniatures. This incipit page shows a large letter *L* constructed of interlaced and spiraling vines of gold leaf, a flight of artistic fancy. The other letters of *Liber* form part of the design. An *I* in silver is slotted among the golden vines. The final three letters appear in gold to the right. The remaining words are written in letters of alternating gold leaf and silver against a densely patterned background of burgundy. This background imitates the expensive silks from Byzantium that Western Europeans admired and regarded as precious objects. Byzantine silks were frequently used to cover such highly valued manuscripts as this Gospel book. TK

INCIPIT SCI EVANGELII.
SCDM. MATHEUM.

BER
GENE
RATIO

7 New Testament
Constantinople, 1133

279 leaves, 22 x 18 cm
($8^{11}/_{16}$ x $7^{1}/_{8}$ in.)
Ms. Ludwig II 4; 83.MB.68

Plate: *Saint Luke,* fol. 69v

The Roman emperor Constantine (the Great) was responsible for two of the most profound acts in European history. As the first emperor to convert to Christianity, Constantine provided official impetus toward the wide-scale spread of the relatively new religion, and when he chose to move the imperial capital away from Rome in 330, he decisively shifted the political and cultural focus of the empire. As the heart of the emerging Byzantine realm, Constantinople (modern Istanbul, located on the Bosporus between Europe and Asia Minor) was considered the "new Rome," and its inhabitants always regarded themselves the true heirs of the classical legacy.

Byzantine art reflects this dual Roman and Christian heritage, as the portrait of Luke in this manuscript demonstrates. The antique garb and careful modeling of the face ultimately stem from classical art, while the placement of the figure against a shimmering gold background suggestive of heaven is consonant with the medieval Byzantine aesthetic. Part of a long tradition of evangelist portraits, the images of Luke and the other three Gospel authors are representative of the twelfth-century Comnenian style (named after the ruling dynastic family). Although based on earlier models from the ninth and tenth centuries, the vigorous drapery and somewhat attenuated poses reveal that Byzantine art was also moving toward a more abstract and dynamic phase.

According to an inscription near the end of the manuscript, this New Testament was finished in the year 1133 by Theoktistos, almost surely in Constantinople, where this scribe wrote another book for a prominent monastery. (However, he is not specifically identified as a monk.) The Getty manuscript is thus one of the few deluxe Byzantine books that can be accurately dated and localized. It serves as a benchmark of the artistic continuity and stylistic innovations in twelfth-century Byzantine art. ASC

ONDITOR
ALME SIDERUM

8 Breviary
Montecassino, 1153

428 leaves, 19.1 x 13.2 cm
(7 9/16 x 5 3/16 in.)
Ms. Ludwig IX 1; 83.ML.97

Plate: Decorated Initial *C*, fol. 259v

This manuscript was made at the Monastery of Montecassino in southern Italy, the cradle of Benedictine monasticism and an important center for the production of books. Among the manuscript's texts is a prayer that names "the Lord's servant Sigenulfus" as the scribe. Undoubtedly a monk of the abbey, Sigenulfus may have been responsible for both the writing and the illumination of this splendid book.

Benedictine monks and nuns lived in organized communities apart from the secular world. Much of their waking day was occupied with the celebration of the eight services that make up the divine office (the prayer liturgy of the Catholic Church, consisting principally of the recitation of psalms and the reading of lessons). Medieval manuscripts containing the texts of the office, called breviaries, were sometimes large volumes intended for communal use, but more often they were small books, like this one, designed to be used by an individual.

The Museum's breviary from Montecassino is extremely richly illuminated, with twenty-eight large decorated initials and over three hundred small initials. This letter *C* formed of panels, interlace, and spiraling tendrils painted in gold and brilliant colors introduces the hymn for the first Sunday in Advent: *Conditor alme siderum . . .* (Creator of the heavens . . .). A pair of bold, blue animal heads form the ends of the letter's curves, and a curious human figure occupies the center of the design. Fantastic doglike creatures twist through the tendrils biting at the vines, each other, and their own bodies. The remainder of the text is in fancy gold capitals. The bright yellow and blue and the biting dogs of the initial are especially characteristic of Montecassino manuscript illumination of the period. ECT

9 Gratian, *Decretum*
Sens or Paris, circa 1170–1180

239 leaves, 44.2 x 29 cm
(17 7/16 x 11 7/16 in.)
Ms. Ludwig XIV 2; 83.MQ.163

Plates: Initial *I* with *Scenes of Secular and Ecclesiastical Justice,* fol. 1
Initial *Q* with *An Abbot Receiving a Child,* fol. 63

As a teacher in Bologna, the monk Gratian organized the study of Church law with his compilation of the *Decretals,* an unprecedented collection of nearly four thousand texts drawn from Early Christian writings, papal pronouncements, and council decrees. Completed sometime between 1139 and 1159 (the year of Gratian's death), the *Decretals* quickly became the standard textbook throughout Europe in the field of canon law. The use of such standardized texts became increasingly important with the formation and rise of universities at the end of the twelfth and throughout the thirteenth centuries, particularly in Paris, Bologna, and Oxford.

In the initial *I* that opens this manuscript, medallions show the king and bishop as representatives of secular and spiritual law, demonstrating the importance of the separation of powers. In the initial *Q,* simony is illustrated as an abbot receives a child into the monastery along with payment from the father. Simony, the improper traffic in holy things, was a significant problem confronted by Church law. Named after Simon Magus, who was reprimanded by Saint Peter for wanting to acquire the power of the Holy Spirit (Acts 8:9–24), it most commonly referred to monetary transactions involved in appointments to Church offices. Abundant medieval decrees indicate that simony was a recurring concern.

With their combination of imaginative hybrid creatures and coiling tendrils, both initials are typical of northern French Romanesque painting strongly influenced by English art. This is evident too in the abbot's robe, where the drapery is rendered in broad patches, revealing the substance of the body beneath the cloth. The decoration of this manuscript connects it to a group of books produced for Thomas Becket, Archbishop of Canterbury, and his secretary Herbert of Bosham while they were in exile in France between 1164 and 1170. It is not clear, however, whether the Getty manuscript and the other books were illuminated in Sens, the site of Becket's exile, or nearby Paris.

ASC

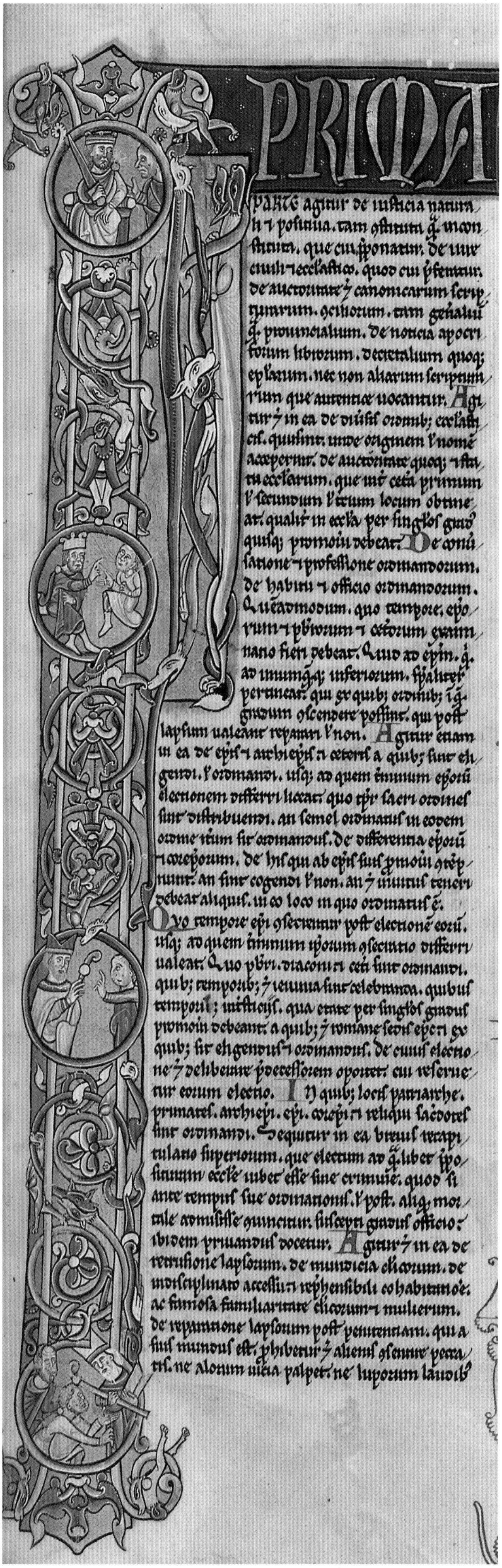
PRIMA

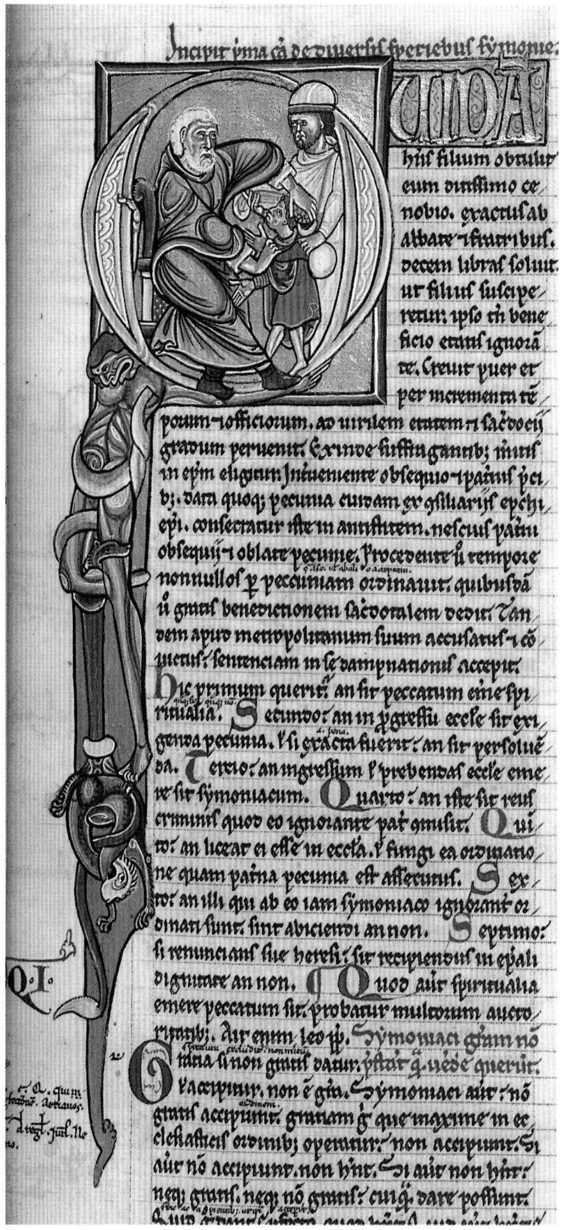
Incipit pma ca de diuersis speciebus symonie.

10 Canon Tables
from the Zeytᶜun Gospels
Hromklay, 1256

8 leaves, 26.5 x 19 cm
(10 7/16 x 7 1/2 in.)
Ms. 59; 94.MB.71

Plates: Tᶜoros Roslin, Canons 2–5, fols. 3v–4

See pages 30–31

Tᶜoros Roslin was the most accomplished master of Armenian manuscript illumination. His work is remarkable both for its consummate artistry and for its incorporation of motifs learned from Western European and Byzantine art. Active in the second half of the thirteenth century, he wrote and illuminated manuscripts for the Cilician royal family and for Catholicos Kostandin I (1221–1267), the highest official of the Armenian Church.

Christianity became the official religion of the Arsacid kingdom of Greater Armenia in the early fourth century. The belief of the Armenian Catholic Church is distinct from the Roman Catholic and Orthodox traditions, although the doctrines of the Armenian Church are similar to those of the Eastern Orthodox faiths. The Armenian language was not a written language until after the adoption of Christianity; the alphabet was most probably created in order to preserve and disseminate scripture, and Bibles and Gospel books number among the most sumptuous of manuscripts in Armenian.

Compiled by Eusebius of Caesarea, canon tables consist of columns of numbers that present a concordance of passages relating the same events in the four Gospels. Canon table pages attracted decoration in manuscript Bibles and Gospel books throughout the Middle Ages, the columns of numbers naturally inviting an architectural treatment. On these pages, Roslin has placed the text within a grand and brilliantly colored architecture with column capitals formed of pairs of birds. The whole shimmers with gold, and the vase at the top of the left page is carefully modeled in silver and gold. The grandeur of the architecture and the symmetry of the trees contrast with the naturalism of the hens that dip their heads to peck at a vine and drink from a fountain. ECT

11 Dyson Perrins Apocalypse
England (probably London), circa 1255–1260

41 leaves, 31.9 x 22.5 cm (12 9/16 x 8 7/8 in.)
Ms. Ludwig III 1; 83.MC.72

Plates: *Unclean Spirits Issuing from the Mouths of the Dragon, the Beast, and the False Prophet* and *The Angel Pouring Out from the Seventh Vessel,* fols. 34v–35

See pages 34–35

Thirteenth-century England saw the creation of a large number of illuminated manuscripts of the Apocalypse (Book of Revelation), Saint John the Divine's vision of the events leading to the Second Coming of Christ at the end of time. The Apocalypse had a particular resonance for Western Europeans in the mid-thirteenth century; recent cataclysmic events, including the invasion of Russia by the Tatars (1237–1240) and the fall of Jerusalem to the Moslems (1244), suggested that the end of time was near. The enigmatic text of the Apocalypse invited interpretation, and this manuscript includes the commentary most commonly found in English Apocalypses, that of Berengaudus (a monk about whom nothing is known except that he wrote this commentary).

Every page of the Dyson Perrins Apocalypse, named for a previous owner of the manuscript, includes a half-page miniature, a brief passage from the Apocalypse (in black ink), and a portion of Berengaudus's commentary (in red ink). The miniatures are in the tinted drawing technique, which reached a level of great sophistication in thirteenth-century England. They vividly illustrate the biblical text in compositions of great clarity. Saint John is often shown experiencing his vision, either from within the scene or peering from the margin through an opening in the miniature's frame.

One miniature (fol. 35) depicts an angel pouring from a vessel, which unleashes "lightnings, voices, thunders, and a great earthquake" (Apocalypse 16:17–18). An oversized Saint John seems to turn back just in time to see the destruction brought about by the earthquake. The "great voice out of the temple from the throne" is represented as a half-length figure of Christ within a mandorla emerging from a building surrounded by clouds. The heavenly temple appears to be suspended from a small peg in the upper margin of the page, a visual delight entirely unaccounted for in the text. ECT

Et vidi de ore draconis & de ore bestie & de ore pseudo prophete spiritus tres inmundos in modum ranarum. Sunt enim spiritus demoniorum facientes signa. Et procedunt ad reges terre totius congregare illos in prelium ad diem magnum dei omnipotentis. Ecce venio sicut fur. Beatus qui vigilat & qui custodit vestimenta sua ne nudus ambulet & videant turpitudinem eius. Et congregavit illos in locum qui vocatur ebraice ermagedon.

Sunt autem spiritus demoniorum facientes signa. vocatione gentium descripta qualiter ad fidem christi venerunt mentionem facit antichristi qui circa finem mundi venturus est. Spiritus vero tres inmundi discipulos designant antichristi qui per universum orbem predicaturi sunt. Qui quoniam homines sunt futuri spiritus inmundi & spiritus demoniorum vocantur quia demones in ipsis habitabunt & per ora eorum loquentur. Qui de ore antichristi & de ore pseudoprophete eius exisse visi sunt: quia per eorum doctrinam filii diaboli efficientur. Qui & de ore draconis exisse visi sunt. quia per os antichristi diabolus loquetur. Ranis videlicet que sunt reptilia inmunda & in luto viventia recte assimulantur. quia sicut rane in sordidis aquis commorantur ita & discipuli antichristi eos facile decipient qui divitiis viciis & sordibus non timuerint sordidari. Clamor & ranarum vox rauca & turpis. impudentissimam eorum predicationem blasphemiis plenam designat. Et procedunt ad reges terre &c. Per reges terre non solum reges sed & populi designantur. Dies autem domini sicut fur in nocte veniet. Cum enim dixerint pax & securitas tunc repentinus eis superveniet interitus.

Et septimus angelus effudit
phialam suam in aerem ⁊
exiuit uox magna de templo
a throno dicens. Factum est.
Et facta sunt fulgura ⁊ uoces ⁊ tonitrua ⁊
terre motus factus est magnus qualis numquam
fuit ex quo homines fuerunt super terram ta-
lis terre motus sic magnus. Et facta est ciui-
tas magna in tres partes ⁊ ciuitates gentium
ceciderunt. Et babilon magna uenit in me-
moriam ante dominum. dare ei calicem uini
indignationis ire eius. Et omnis insula fu-
git ⁊ montes non sunt inuenti. Et grando
magna sicut talentum descendit de celo in homines ⁊ blasphemauerunt deum
homines propter plagam grandinis quoniam
magna facta est uehementer.

Per septimum istum angelum predicatores sancti
qui temporibus antichristi fuerunt designantur.
Angelus qui phialam suam in aerem effudit quia predicatores sancti uanitatem ⁊ impiis hominibus quod pro eterna
pena sunt dampnandi denuntiabunt. Et exiuit
uox magna ⁊c. Vox magna uox est predica-
torum sanctorum. Per templum ecclesia intelligitur. A templo
ergo uox exiit quia ab ecclesia uox sancte predicationis
procedit. Que et a throno exisse dicitur quia ecclesia dei thro-
nus est dei ⁊ in illa sedens requiescat. Quid autem
hec uox dicat subdendo manifestat. Factum est.
id est finis mundi instat in quo omnia que predicta
sunt a domino ⁊ a sanctis complebuntur. Per fulgura uero
miracula que per sanctos suos facturi sunt designantur. Legimus namque in superioribus heliam
⁊ enoch plurima signa esse facturos. Per uoces
uero predicatio sanctorum. Per tonitrua autem terror ig-
nis exprimuntur.

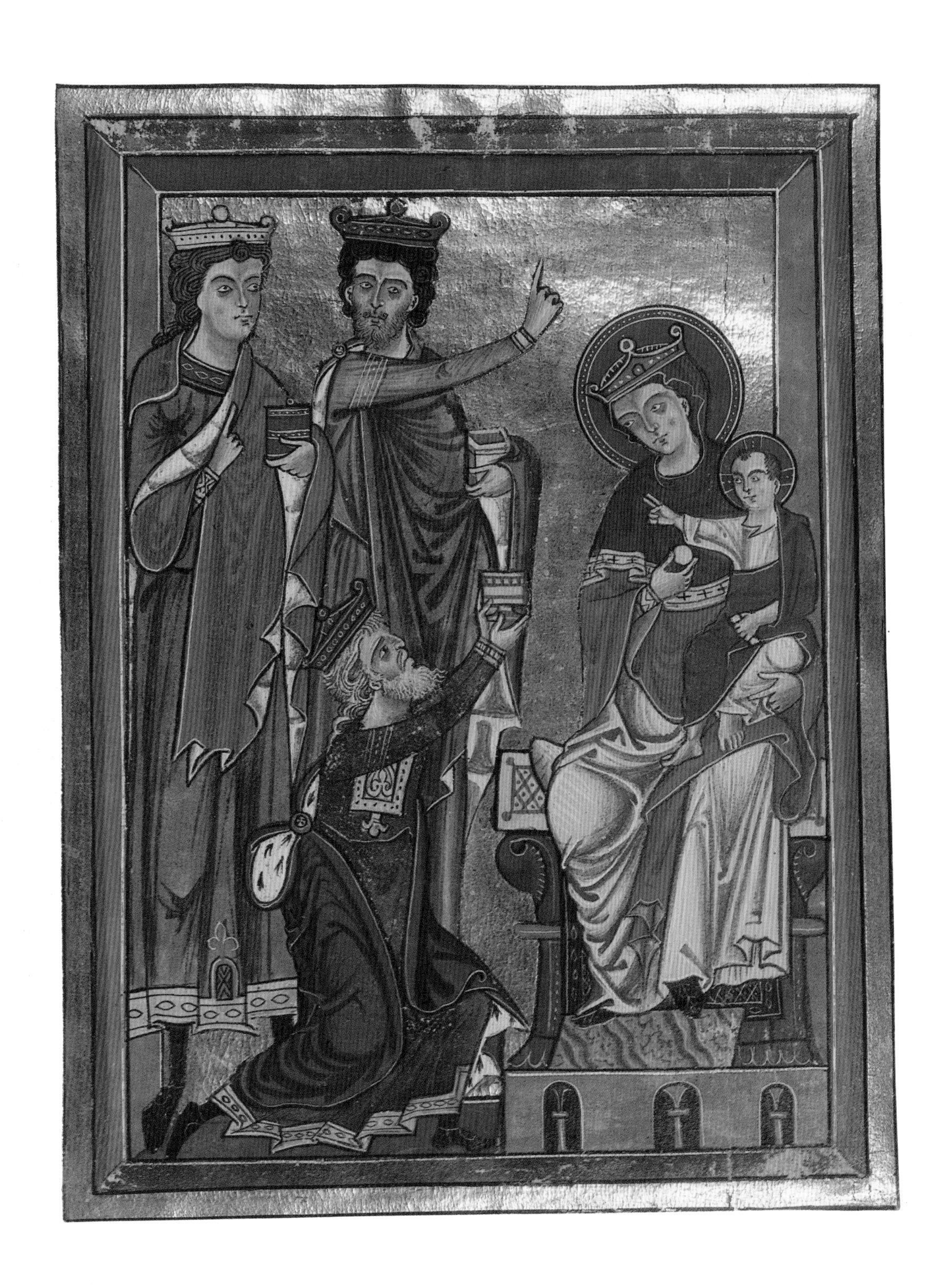

12 Two Miniatures from a Psalter
Würzburg, circa 1240

17.7 x 13.6 cm (7 x 5 5/16 in.)
Ms. 4; 84.ML.84

Plate: *The Adoration of the Magi,* leaf 2

Situated at the heart of monastic religious life, the recitation of the psalms played a central role in Christian devotions throughout the Middle Ages. By the thirteenth century, the psalms became the focus of private devotion. A psalter consists of all 150 psalms along with a calendar of Church feasts and other texts. It was the first important prayer book for lay worshipers and a vehicle for lavish decoration.

This *Adoration of the Magi* is one of two large miniatures in the Getty collection from a picture cycle removed from a thirteenth-century psalter made at Würzburg in Bavaria. Eighteen others in this cycle are known, including sixteen in the British Library, and the suite of miniatures was undoubtedly larger still. (The rest of the manuscript—including its text—is lost.) The miniatures tell the story of the Life of Christ, beginning with the Annunciation to the Virgin Mary, through Christ's infancy, trials, death, and Resurrection. This dramatic sequence of New Testament miniatures would have preceded the psalms themselves and focused the worshiper's attention on the heart of Christianity—the example of Christ himself.

The Würzburg school of manuscript illumination flourished in the middle of the thirteenth century. Our knowledge of it derives from this fragmentary psalter and half a dozen other books that survive, most of them also psalters (see no. 13). Whereas the finest painted books throughout the Middle Ages feature expensive pigments and precious metals, the backgrounds of highly burnished gold leaf are especially characteristic of German, French, and Flemish manuscripts of the thirteenth century. Lacking any indications of setting, the luminous, undifferentiated background focuses the viewer's attention on the story of the three kings from the east, who follow a star in search of the baby Jesus, "he who is born to be king of the Jews." The king at the center, with arm raised, points to the unseen celestial body that led him and his companions to Bethlehem. The artist depicts the Christ Child not in the humble manger where he was born but sitting prominently in the lap of his mother, who is seated on a regal throne. TK

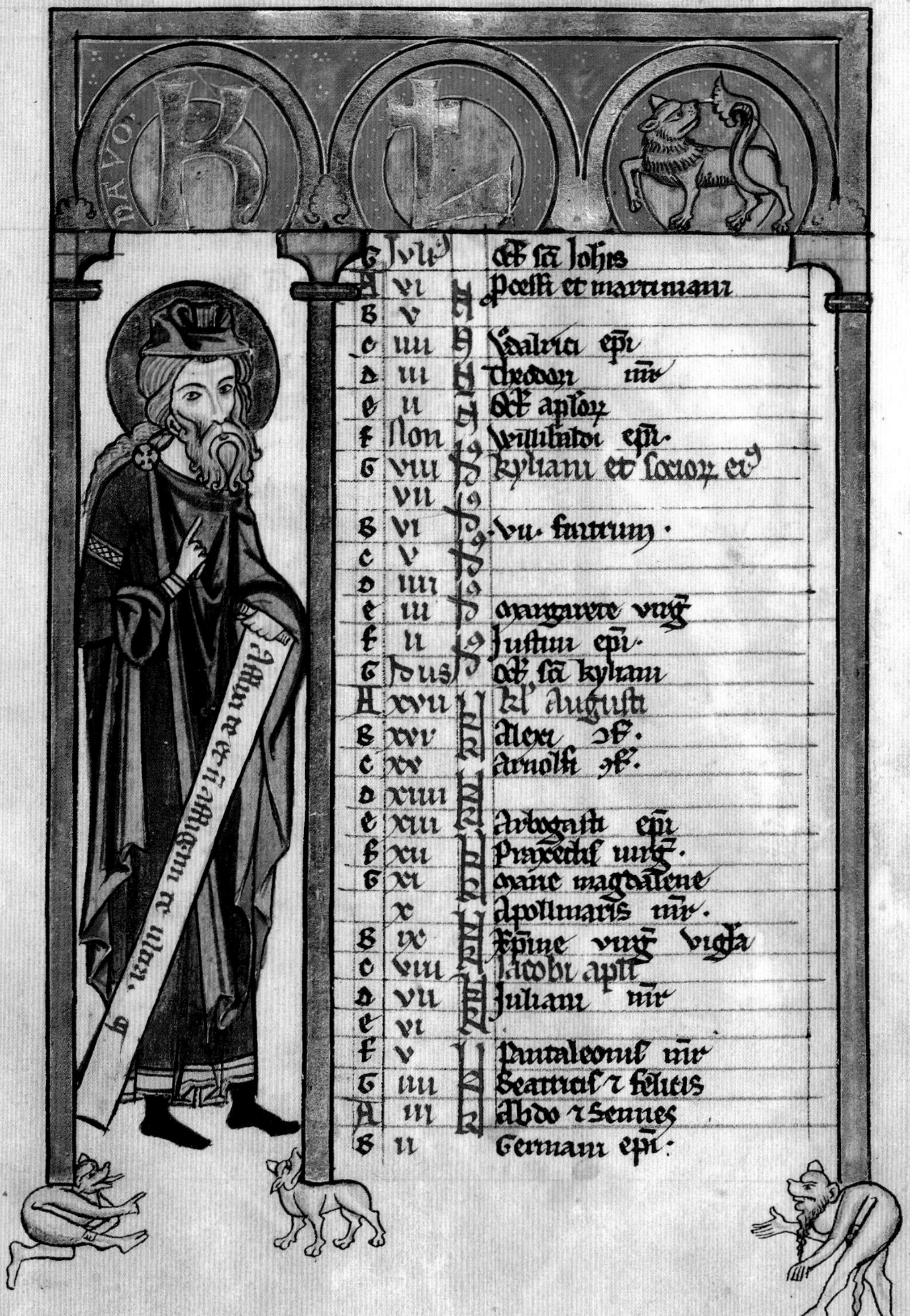

c	Iulij	Oct. sci Johis
A	vi	Processi et martiniani
b	v	
c	iiii	Udalrici epi
d	iii	Theodori mr
e	ii	Oct. aplor.
f	Non.	Willibaldi epi.
g	viii	Kyliani et socior. eius
	vii	
b	vi	vii. fratrum.
c	v	
d	iiii	
e	iii	Margarete virg.
f	ii	Iustini epi.
g	Idus	Oct. sci kyliani
A	xvii	Kl. augusti
b	xvi	Alexi cf.
c	xv	Arnolfi cf.
d	xiiii	
e	xiii	Arbogasti epi
f	xii	Praxedis virg.
g	xi	Marie magdalene
	x	Apollinaris mr.
b	ix	Xpine virg. vigilia
c	viii	Iacobi apli
d	vii	Iuliani mr
e	vi	
f	v	Pantaleonis mr
g	iiii	Beatricis et felicis
A	iii	Abdon et sennes
b	ii	Germani epi.

Deus qm̄ intrauerunt aque usq; ad a
nimam meam I n fixus sum ĩ limo

13 Psalter
Würzburg, circa 1240–1250

192 leaves, 22.6 x 15.7 cm
(8 15/16 x 6 3/16 in.)
Ms. Ludwig VIII 2; 83.MK.93

Plates: Calendar Page for July, fol. 4
Decorated Initial *S* with *Griffin and Rider,* fol. 76

See pages 38–39

During the course of the thirteenth century the illuminated psalter, especially in Germany, Flanders, France, and England, became the most profusely decorated of books. This psalter was made in Würzburg toward the middle of the thirteenth century by artists closely related to the painter of the preceding miniature. They must have known each other and on occasion worked together. This book was decorated from front to back with a range of both religious subjects and playful decoration.

The book opens with a calendar listing the saints' days and other holidays celebrated in the course of the Church year. The calendar is illustrated with images of lesser prophets of the Old Testament, Nahum appearing for July (fol. 4). He holds a scroll from his writings: "Though I have afflicted thee, I will afflict thee no more." (Nahum 1:12).

In order to facilitate their recitation during the course of the week's devotions, the psalms are divided into a total of ten sections. Full-page miniatures including both Old and New Testament subjects appear before Psalm 1, and others precede Psalm 51 and Psalm 101. The illuminator introduced the seven remaining sections with large decorated and inhabited initials, the latter being particularly inventive. In the example shown here the initial *S* has been transformed into a griffin ridden by a loosely robed youth and entwined with foliage and other beasts. The text *Salvum me fac* begins Psalm 68 (Save me, oh Lord, for the waters threaten my life . . .).

Often the names of illuminators and patrons of even the finest medieval manuscripts are not known. Artists in particular rarely signed their works. We know this book was made in Würzburg in part because of the liturgical indications in the text and in part because of its close relationship to the illumination of a Bible made in Würzburg in 1246. One of the painters of the Bible signed one of its miniatures: *Hainricus pictor* (Henry the Painter). The illuminators of the Getty psalter and of the miniatures in no. 12 undoubtedly knew Henry, but their names remain lost to us.

TK

14 Wenceslaus Psalter
Paris, circa 1250–1260

203 leaves, 19.2 x 13.4 cm
(7 9/16 x 5 1/4 in.)
Ms. Ludwig VIII 4; 83.MK.95

Plate: Initial *B* with *David Playing Before King Saul and David Slaying Goliath,* fol. 28v

See page 42

The mythic place that Paris occupies in the modern imagination as a center of beauty and the visual arts has deep historical roots. They penetrate to the twelfth century when the Gothic style emerged in architecture and art of the Île-de-France, the region along the River Seine with Paris at its center. Monumental stained glass enriched the walls of its cathedral, churches, and chapels. In the thirteenth century a lively industry of book production flourished there as well. The city quickly became famous throughout Europe as a center of manuscript painting. Indeed, in the early fourteenth century in distant Florence, the poet Dante (1265–1321) mentions in *The Divine Comedy* "the art which in Paris is called illuminating."

This psalter offers evidence of the international appeal of Parisian Gothic manuscript illumination. It contains over 160 narrative scenes from the Old and New Testaments and countless initials painted with generous quantities of gold leaf and costly pigments. Within a generation of its creation, a Bohemian nobleman (in the modern-day Czech Republic) acquired it. Some scholars believe that he was no less than King Wenceslaus III of Bohemia (r. 1305–1306).

The most important decoration of a psalter is the *Beatus* initial page, containing the illustration to the first psalm: *Beatus vir* . . . (Blessed is the man . . .). The initial is formed by vines that culminate in animal heads and roundels filled with stories of David. In the *B*'s upper lobe the young David plays his harp before Saul; in the lower one the boy slays Goliath. As if with jewels, the frame of the page is encrusted with additional scenes from David's life. The crowded design of this initial is not unlike that of stained-glass windows, constructed of a pattern of lozenges and roundels, each with an individual scene, usually narrated with only a few figures. Whereas the luminosity of stained glass derives from external light transmitted through the colored glass, in Gothic books the highly burnished and reflective backgrounds of gold leaf next to saturated colors strive for a similarly brilliant effect. TK

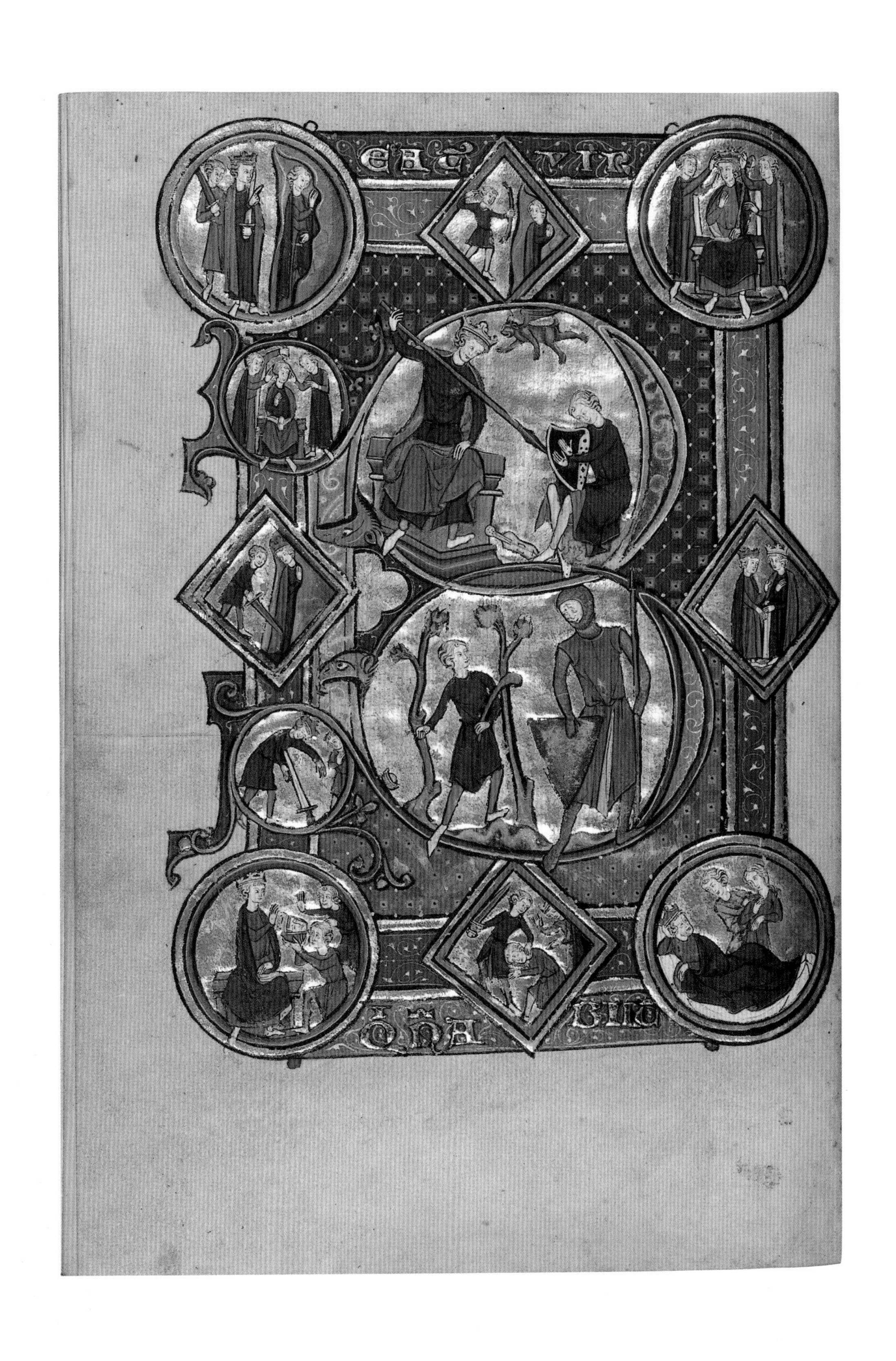

15 Bute Psalter
Northeastern France,
circa 1270–1280

346 leaves, 16.9 x 11.9 cm
(6 11/16 x 4 11/16 in.)
Ms. 46; 92.MK.92

Plate: Bute Master, Initial *D* with *King David Pointing to His Mouth,* fol. 52v

This psalter, formerly in the collection of the Marquess of Bute in Scotland, gives the name "Bute Master" to its anonymous artist. The Bute Master worked in the prosperous cities of the Franco-Flemish border region, contributing to the illumination of a dozen sacred and secular manuscripts. He or she sometimes collaborated with other illuminators, a common practice in the thirteenth century, but was entirely responsible for this manuscript's 190 historiated initials.

An intimate connection between text and image is evident in some of this book's initials. One of the largest introduces Psalm 38 (fol. 52v), the first of fourteen psalms recited in the pre-dawn prayer service of Matins on Tuesdays. The subject of the initial

Terra autem erat inanis et
uacua: et tenebre erant super
faciem abyssi: et sps di fereba
tur sup aquas. Dixitque ds.
Fiat lux. Et facta est lux. Et
uidit ds lucem quod esset
bona: et diuisit lucem ac te
nebras. Appellauitque lucem
diem: et tenebras noctem. Fa
ctumque est uespe et mane: di
es unus. Dixit quoque ds. Fiat
firmamentum in medio a
quarum: et diuidat aquas
ab aquis. Et fecit ds firma
mentum: diuisitque aquas
que erant sub firmamento.
ab his que erant sup firma
mentum. Et factum est ita.
Vocauitque ds firmamentū
celum. Et factum est uespe et
mane: dies secundus. Dixit
uero ds. Congregentur aque
que sub celo sunt in locū unū:
et appareat arida. Factūque ē
ita. Et uocauit ds aridam
terram: congregationesque
aquarum appellauit maria.
Et uidit ds quod esset bonū:
et ait. Germinet terra herbā
uirentem. et facientē semen:

was chosen according to the word (*ad verbum*) of the psalm, which opens "I said: I will take heed to my ways that I not sin with my tongue." King David, purported author of the psalms, points to his mouth with his right hand, a direct visualization of his promise to avoid sinning with his tongue. It is less clear why David points to the ground with his other hand. Perhaps this gesture alludes to the psalmist's eventual burial, for the psalm refers to the "numbering of the days" and the "passing" of its author.

The scene within the initial is complemented in the *bas-de-page* (literally, "bottom of the page"), where a seated woman points to a soldier who looks back at her as he points over to the facing page. The glances and gestures of all the figures, together with the vignette of the dog chasing a hare in the upper margin, lead the eye around the page, infusing the ensemble with an energy that undoubtedly pleased the manuscript's thirteenth-century aristocratic owner as much as it does the twentieth-century museum visitor. ECT

16 Marquette Bible
Probably Lille, circa 1270

3 volumes, 273 leaves,
47 x 32.2 cm (18½ x 12$^{11}/_{16}$ in.)
Ms. Ludwig I 8; 83.MA.57

Plate: Initial *I* with *Scenes of the Creation of the World and the Crucifixion,* vol. 1, fol. 10v

The Bible, understood to be the written word of God, is the central holy book of Christianity. Comprising Jewish sacred writings, the four Gospel accounts of Jesus' life, the letters of Saint Paul, and other texts, it is a very long book indeed. Manuscripts of the Bible were generally multivolume, large-format books designed for use at a lectern until the rise of the universities created a demand among students for small-format, portable Bibles. At around the same time, the writing and embellishment of Bibles became less and less the work of monks and more and more the activity of lay artisans.

In thirteenth-century France, the traditional large-scale format was retained at the same time that "pocket Bibles" were mass-produced in the university city of Paris. The Museum's Marquette Bible is one of several artistically related lectern Bibles made for religious institutions in northeastern France and illuminated by teams of lay artists. The Marquette Bible's illumination takes the form of historiated initials. Originally, the Bible must have had around 150 painted initials (of which 45 survive). It is hardly surprising, given the size of the undertaking, that scholars have identified the work of six different artists among the surviving initials, and we can well imagine that the team of illuminators responsible for the original seven-volume Bible was larger still.

The main artist of the Marquette Bible painted most of the initials in the early part of the text, including the glorious Genesis initial (vol. 1, fol. 10v). This initial introduces not only the book of Genesis but also the Bible as a whole; the series of scenes of Creation (as told in Genesis) ends with the New Testament scene of the Crucifixion. This combination highlights the Christian belief that Christ's death restored the world's communion with God, lost when Adam disobeyed him by eating the fruit of the Tree of Knowledge in the Garden of Eden. ECT

Incipiunt
hore de be
virgine.
Domine la
bia mea a
peries: Et os
meum an
nuntiabit laudem tuam.
Deus in adiutorium meum
intende domine ad adiu
vandum me festina.
Gloria patri et filio et spiritui sancto.
Sicut erat in principio et nunc
et semper et in secula seculorum amen.

17 Ruskin Hours
Northeastern France, circa 1300

128 leaves, 26.4 x 18.4 cm
(10 3/8 x 7 1/4 in.)
Ms. Ludwig IX 3; 83.ML.99

Plate: Initial *D* with *The Annunciation,* fol. 37v

By the fourteenth century, the book of hours replaced the psalter as the most important text for the daily personal devotions of the Christian faithful. It takes its name from the Hours of the Virgin, the book's core text. These prayers are organized for private recitation at the canonical hours, eight appointed times of the Church day. During the later Middle Ages in particular the Church encouraged the growth of private prayer and meditation among the laity. The rise in this practice and the expansion of wealth led to a demand on the part of the aristocracy and the burgeoning merchant class for fancy, decorated prayer books. Northern France was one of the prosperous regions where prayer books flourished. Not only Parisian workshops but others located throughout the north in towns like Lille, Cambrai, and Douai profited from the demand for devotional books.

Following the traditional iconography of the Hours of the Virgin, the illuminator of this large prayer book has illustrated each of the eight hours with an episode from Mary's life. For Matins, the first hour, he has depicted the Annunciation to the Virgin Mary inside the initial. All the decoration on this page springs from the large *D* in winding, spiraling, and elongated vines, an exuberant visual complement to the text's joyful opening words taken from Psalm 50: "O Lord, open my lips, and my mouth shall proclaim your praise." The smaller initial *D* shows a devout young man dressed in a simple tunic, raising his eyes in prayer to a receptive God. Figures in prayer offer similar models of devotion throughout the book's borders and smaller initials.

The jousting soldiers in the border reflect a popular aristocratic pastime of the day. Such marginal figures, clearly motifs to charm and amuse the viewer, occasionally appear to comment, sometimes humorously, on devout themes. Often, as is the case here, their relationship to the book's central religious imagery is not obvious.

This book belonged to the influential English art critic John Ruskin (1819–1900), who delighted above all in the rhythmic extenders of the book's historiated initials. He extolled them as "bold" and "noble." TK

corporis magnitudine.
Nam circa semitas per quas
elephantes solito more gra-
diuntur delitescens colla
eorum suis nodis alligat

ac suffocatos perimit.
Gignitur in ethiopia
et in india. in ipso ien-
dio iugis estus

De belua que dicitur cetus.

18 Bestiary
Flemish, circa 1270

102 leaves, 19 x 14.4 cm
(7 1/2 x 5 5/8 in.)
Ms. Ludwig XV 3; 83.MR.173

Plate: *Two Fishermen, Believing Themselves at an Island, Make Their Camp on the Back of a Sea Creature,* fol. 89v

The bestiary, or "book of beasts," was one of the most popular books in the twelfth and thirteenth centuries, when its text was increasingly expanded, translated into various vernacular languages, and profusely illustrated. This allegorical interpretation of real and imaginary animals was based principally on the *Physiologus,* a Greek text written in the first centuries of the Christian era and translated into Latin in the fourth century.

From the start, such works were not scientific in the modern sense—they were more interested in drawing moral lessons than in providing objective investigation. While embracing the philosophy that the observation of the physical world leads to an understanding of heavenly operations, the *Physiologus* innovatively imbued pagan material with new Christian interpretations. The bestiary that took form at the end of the twelfth century incorporated many other early medieval sources into its text, above all material from the seventh-century encyclopedia of Bishop Isidore of Seville.

The treatment of the large sea creature called the *aspidochelone* is typical. One characteristic of the animal is that it floats with its huge back emerging above the waves, remaining motionless for long periods of time. After sand has settled there and vegetation has grown, sailors mistake the beast for an island and beach their ships on it. When the sailors light their campfires, the monster feels the heat and plunges suddenly into the watery depths. In this miniature, the artist succinctly captures all the dramatic potential of the story. While the sailors react in distress to the great beast's dive, one victim tumbles over to certain death; the fate of the man clinging tenuously to the boat hangs in the balance.

The *aspidochelone* is understood allegorically as the wily devil who deceives sinners, plunging them into the fires of hell. Similarly, the little fish that swim into the creature's mouth, attracted by the sweetness of its breath, are understood as those who are easily tempted and so swallowed by the devil. This kind of moralizing was standard in the bestiary and related texts, many of which were written by and for monks. ASC

19 Antiphonal
Bologna, late thirteenth century

243 leaves, 58.2 x 40.2 cm
($22^{15}/_{16}$ x $15^{13}/_{16}$ in.)
Ms. Ludwig VI 6; 83.MH.89

Plate: Master of Gerona, Initial *A* with *Christ in Majesty,* fol. 2

Splendidly illuminated choir books, large enough to be seen by a group of singers, stood open on lecterns in Christian churches throughout Western Europe during the High Middle Ages and Renaissance. The two principal types of choir book are the antiphonal and the gradual. An antiphonal contains the chants of the divine office—the eight prayer services celebrated daily by monks, nuns, and clerics of the Catholic Church. The musically elaborate portions of the mass—the Christian rite in which bread and wine are consecrated and shared—are found in a gradual.

The illumination of choir books primarily takes the form of historiated initials. The first and most impressive initial in this antiphonal is an *A* with *Christ in Majesty* (fol. 2). Its subject was inspired by the chant it introduces, which relates that the speaker "sees the coming power of the Lord." This "coming power" is understood in a Christian context as the return of Christ at the end of time, when he will sit in judgment of all humanity. The prophet Isaiah (whose words provided the inspiration for the text of the chant) "sees" Christ from the roundel at the lower left.

The illuminator of this antiphonal was well versed in the most recent trends in panel painting. His style resembles that of the Florentine painter Cimabue (circa 1240–1302[?]), who was described by the first historian of Italian art, Giorgio Vasari (1511–1574), as the *prima luce* (first light) of painting. Vasari thus saw Cimabue as standing at the beginning of a new development in Italian art that culminated in the work of the High Renaissance artist Michelangelo. Like Cimabue, the Master of Gerona was profoundly influenced by Byzantine icon painting but also made great strides in naturalistic representation, as evidenced in this spatially ambitious composition of Christ enthroned with standing angels. ECT

appropinquauit. ps. Beat'
R. Ex syon speties decoris eius.
Deus noster manifeste ueniet.
uir. euouae.
R.
spitiens
alon ge
ecce

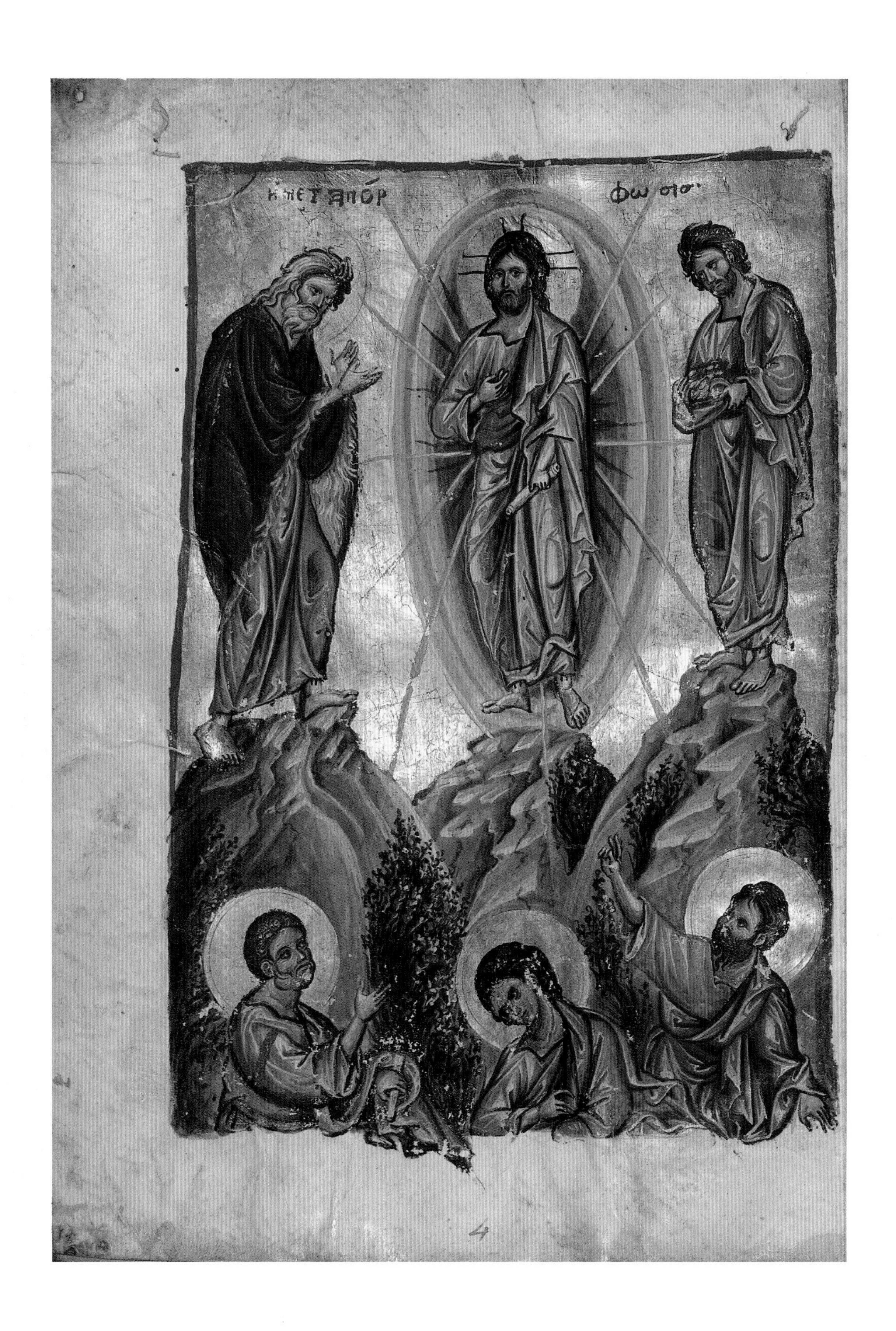

20 Gospel Book
Nicaea or Nicomedia,
early and late thirteenth century

241 leaves, 20.5 x 15 cm
(8⅛ x 5⅞ in.)
Ms. Ludwig II 5; 83.MB.69

Plate: *The Transfiguration,* fol. 45v

With the sack of Constantinople in 1204 by Crusaders from Western Europe, Byzantine political administration shifted away from the imperial capital now dominated by the invaders. Based on artistic and paleographic comparisons with other manuscripts, the Getty Museum's Gospel book can be dated to this critical moment in European history. The specific place of its origin has not been determined; Nicaea (modern Iznik), Nicomedia (Izmit)—both not far from Constantinople—and Cyprus have all been suggested as possibilities. The manuscript is thus an important witness to the continued artistic production in the Byzantine provinces at a time of political disruption.

This *Tetraevangelion* (the Greek term for a Gospel book) contains nineteen full-page illuminations: four evangelist portraits and fifteen images illustrating various key feast days in the Christian calendar. Only the evangelist images and two feast pictures can be dated to the beginning of the thirteenth century, however, while the other thirteen images were painted toward the end of the same century. These later pages were inserted as replacements for a portion of the earlier cycle that had deteriorated over the course of the 1200s. Byzantine manuscript painters often coated the bare parchment with egg white, which originally had the effect of giving the page a slick and glossy appearance, but which also led to extensive paint flaking in the miniature. The problem was sufficiently common that Planudes, head of a monastic scriptorium, wrote in a letter of 1295:

> For if the parchment leaves should somehow see water, the writing on them erupts and quakes with the egg, and the work of the scribe turns out into thin air, clean gone.

The miniature of the Transfiguration is representative of the later Palaeologan style that flourished after the Crusaders were expelled in 1261. Named after the imperial family and lasting well into the fifteenth century, this style features large-scale figures based on earlier Byzantine models enlivened by dramatic gestures and an intensity of feeling. *The Transfiguration* and the other Palaeologan images can be dated to around 1285, but, like the earlier miniatures of the book, the place of their production remains uncertain.

ASC

21 Two Miniatures from a Book of Old Testament Prophets
Sicily, circa 1300

7.3 x 17.4 cm (2⅞ x 6⅞ in.)
Ms. 35; 88.MS.125

Plate: *The Vision of Zechariah*, leaf 2

Since the Renaissance some collectors have prized older illuminated manuscripts more for their decoration than for their texts. Thus at a time when bibliophiles still actively commissioned new illuminated manuscripts, other collectors would cut the miniatures and other painted decoration from older books. The practice continued for centuries. In the late eighteenth century the Basel art dealer Pieter Birmann assembled an album of 475 cuttings from scores of medieval manuscripts of all types. The Getty Museum owns two miniatures from that album; the other represents *The Assassination of Sennacherib*. They probably derive from a book of Old Testament prophets.

The rare subject illustrated here is the first of Zechariah's eight visions. A translation from the Latin Vulgate Bible, which was the likely source for the illuminator, reads:

> I saw by night, and behold a man riding upon a red horse, and he stood among the myrtle trees that were in the bottom and behind him were horses, red, speckled, and white. And I said: "What are these, Lord?" and the angel that spoke in me said to me: "I will show you what these are:" And the man that stood among the myrtle trees answered, and said: "These are they, whom the Lord has sent to walk through the earth." And they answered the angel of the Lord, that stood among the myrtle trees and said: "We have walked through the earth and behold all the earth is inhabited and at rest."
>
> (Zechariah 1:8–11)

The artist departs from the mystical text by showing the man in the vision mounting one of the horses rather than simply standing among the myrtles, by showing one red horse instead of two, and by showing the angel at Zechariah's side.

The elongated proportions of the figures and their small heads are particularly characteristic of this moment in Byzantine art. Textual and paleographic evidence, however, suggests that the illuminator, even though he was probably Greek, painted these miniatures in a book written in Western Europe. Such an artist would have resided most likely within the Greek communities of Sicily. TK

puestos entre el cō
prador et el uende
dor.
De nouatione ātous
uoanto i como se
deue nōpnar el ac
De iure emphi tor.
teotico i ɔditōne
ualis diu l'directi.
Del dreito del ɔtrac
to pduráble i dela
ɔdition del sēynno
rio del puesto o del
sēynnorio ppio.
De fide iussoribꝫ
De los fiatores
De fed iobꝫ fide iusso
rum l malefitiorꝫ.
De los hechos delos fia
tores o delos mal
feitores.
De donationibꝫ
De las donationes.
De absolutionibꝫ
De las pagas.

Incipit liber qntꝫ.
de rebus creditis.
Es assaber. delas
cosas que a credē
el uno al otro.

la auer: la quoal
el creedor demā
dar nō sera tenido
de tornar batailla
si aquello que niegꝫ
no es mas de .x. ss.
et esto niega iurādo.
Por reffrer et coer
enar la maleza
de los ācreedores q
muytas uezes deq
enē los iustos. i c.

Este capitulo fau
laras cabo el comiē
ço del tercero libro. to
do ɔpuito. en el titlo
de pscriptionibꝫ. de
Establescemos
que daqui deys
adelant ninguna
deuda nō sea demā
dada sin publico i
strumēto. si nō fuere
puado por bonos tes
tigos q homenage i
fue feito sobre aquella

22 *Vidal Mayor*
Northeastern Spain,
circa 1290–1310

277 leaves, 36.5 x 24 cm
(14 3/8 x 9 7/16 in.)
Ms. Ludwig XIV 6; 83.MQ.165

Plate: Initial *E* with *An Equestrian Duel Between a Creditor and Debtor*, fol. 169v

In 1247, with the reconquest of Spain from the Moslems virtually complete, King James I of Aragon and Catalonia (r. 1214–1276) determined to establish a new systematic code of law. He entrusted the task to one of the leading court figures, Vidal de Canellas, Bishop of Huesca, who had studied law in the famous University at Bologna. Vidal formulated two versions in Latin, and the larger is commonly called the *Vidal Mayor*.

The original Latin *Vidal Mayor* no longer exists, and the Getty Museum's manuscript is the only known copy of the code, preserved in a vernacular Navarro-Aragonese translation. It is thus a critical document of the laws and feudal customs of Aragon. Of particular interest are the cases that deal with Moslems and Jews, as well as with the different classes of Christian society. The *Vidal Mayor* shows clearly, in word and image, that the king's law was applicable to all the inhabitants of the realm.

The historiated initial that opens Book 5 suggests this historical context of the *Vidal Mayor*. This section deals with issues of credit, and the scene depicts a dispute and ensuing duel between a creditor and debtor in the presence of the king. The prominently displayed heraldic devices seem to indicate that the contest is between a Christian and a Moor. The crescent would have called to mind a symbol of the Spanish Moslems, though in this manuscript it may simply be used as a reference to a "foreigner."

With 156 historiated initials, the *Vidal Mayor* is unsurpassed in early fourteenth-century Spanish book illumination. The distinctive style of the figures, the predominance of red, blue, and gold, and the types of animals and beasts used to embellish the initials are all elements of French Gothic art (nos. 14–17). The manuscript was likely produced in one of the major urban centers of northeastern Spain, perhaps Barcelona or Pamplona, by a French artist or perhaps by one trained in Paris or northern France. The scribe of the book, Michael Lupi de Çandiu, may also have been responsible for the translation of the text.

ASC

23 *Vita beatae Hedwigis*
Silesia, 1353

204 leaves, 34.1 x 24.8 cm
(13 7/16 x 9 3/4 in.)
Ms. Ludwig XI 7; 83.MN.126

Plate: *Saint Hedwig of Silesia Adored by Duke Ludwig of Legnica (Liegnitz) and Brzeg (Brieg) and Duchess Agnes,* fol. 12v

The *Life of the Blessed Hedwig* manuscript is a key monument of Central European painting in the fourteenth century. It is the earliest extant illustrated account of the holy Silesian noblewoman Hedwig, who lived from 1174 to 1243 and was canonized in 1267, a remarkably short time after her death. The text and interspersed illuminations reveal much not only about Hedwig's life but also about female spirituality in the High Middle Ages. Unlike early Christian saints, who were typically chaste martyrs, saintly women of the later medieval period were often devoted wives and mothers. Hedwig's deeds, focusing on intense prayer, physical mortification, and extraordinary acts of charity, illustrate various channels used by medieval women to relate spiritually to Christ.

The frontispiece portrays the richly attired saint as a widow with attributes relating to her holy life: the statuette of Mary refers to Hedwig's devotion to the Virgin, the book and rosary to her numerous prayers, and her bare feet to an ascetic existence. The execution of the page reflects the latest style in Bohemian painting, which flourished in the mid-fourteenth century under the Holy Roman Emperor Charles IV. The gently curving figure of Hedwig is vigorously modeled and painted in a manner reminiscent of Central European polychrome sculpture, which also recalls the elegance of contemporary French Gothic art.

The saint stands before her throne, towering over the adoring Duke Ludwig and Duchess Agnes, who commissioned this manuscript. Ludwig, a fifth-generation descendant of Hedwig, was politically a relatively minor Silesian duke but an ambitious patron of building and artistic programs. The manuscript was intended as a monument to the duke's glorious family history and was originally destined for the nunnery in Legnica founded by the saint herself. According to Ludwig's will of 1396 (two years before his death), the codex was sent instead to the so-called Hedwig Convent in Brzeg that the duke had established. The book's text and illustrations would have provided the nuns with a model for their own behavior. ASC

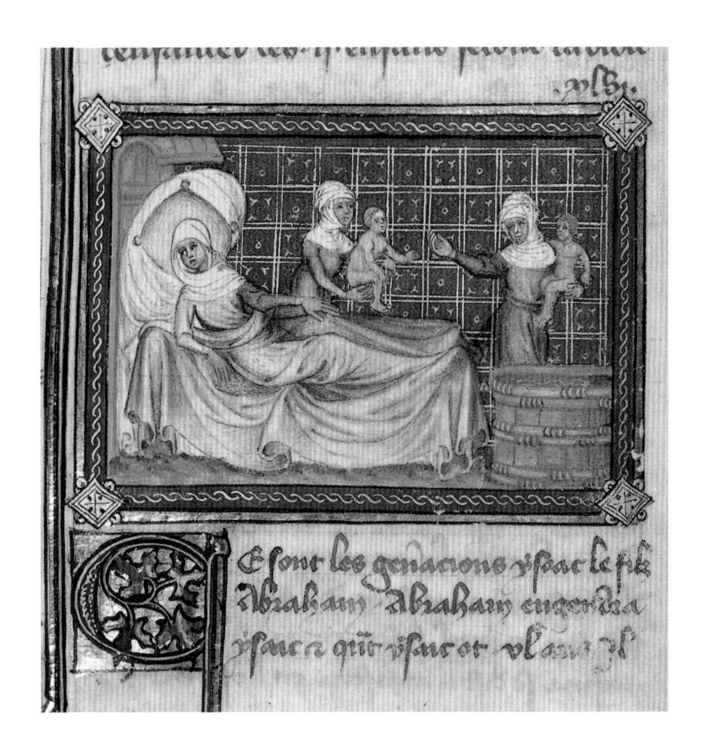

24 Guiart des Moulins,
Bible historiale
Paris, circa 1360–1370

2 volumes, 608 leaves, 35 x 26 cm
(13¾ x 10¼ in.)
Ms. 1; 84.MA.40

Plates: Master of Jean de Mandeville, *The Birth of Esau and Jacob,* vol. 1, fol. 29v
Joseph in the Well, vol. 1, fol. 39
King David with Musical Instruments, vol. 1, fol. 273
The Fool and a Demon, vol. 1, fol. 284

The Bible in its entirety did not become widely accessible in the vernacular until the fourteenth century. In France it was known largely through an extravagantly adulterated version called the *Historical Bible.* Compiled at the end of the thirteenth century by the cleric Guiart des Moulins, it began as a translation of the *Historia scholastica* (Scholastic History) written in Latin by another Frenchman, Peter Comestor (circa 1100–1179). Peter's book stressed the role of scripture as a record of historical events. It consists of his commentary upon excerpts from the Bible. To his translation of the *Scholastic History,* Guiart added further commentary and translations of complete books of the Bible. Even before Guiart's death (by 1322) his book began to appear in an expanded version, supplemented by French translations of all the Bible's books and some of the apocrypha that he had not translated. In the end it had grown to resemble a complete Bible with the addition of commentaries, apocryphal writings, and devotional texts. Like Peter, Guiart emphasized the historical narrative.

The distinctive technique of painting called grisaille (literally "gray" or "painting in tones of gray") enjoyed as much popularity in fourteenth-century France as the

Historical Bible. In this manuscript the costumes of the figures are painted gray while faces and hands are rendered in flesh tones and touches of color. Found throughout French books of this era, the technique attracted many illuminators during the successive reigns of King John the Good (r. 1350–1364) and Charles V (r. 1364–1380) in particular. The brightly patterned backgrounds of the miniatures underscore the three-dimensionality of the delicately drawn, palely colored figures.

Bibles and *Bibles historiales* were often embellished with scores of illuminations. The Getty two-volume example has seventy-three miniatures, their subjects mostly taken from the Old Testament. The first two shown here illustrate scenes from Genesis, the birth of Esau and Jacob and Joseph tossed by his brothers into a well. The next two illustrate the psalms. Artists often introduced the psalms with the scene of King David playing his harp. The fool taunted by a demon illustrates Psalm 52, which begins: "The fool says in his heart 'There is no God.'" TK

25 Missal
Bologna, between 1389 and 1404

277 leaves, 33 x 24 cm (13 x 9 7/16 in.)
Ms. 34; 88.MG.71

Plate: Master of the Brussels Initials, *The Calling of Saints Peter and Andrew,* Initial *D* with *Saint Andrew,* and Initial *Q* with *Saint Peter,* fol. 172a

A missal contains the texts of the mass, which has as its focus the celebration of Holy Communion. The book has several sections. Masses celebrated on Sundays and on feast days commemorating events in the Life of Christ are collected in the Proper of Time (*temporale*). The feast days for individual saints are collected in the Proper of Saints (*sanctorale*). The latter opens with the feast of Saint Andrew (November 30) and is illustrated here by *The Calling of Saints Peter and Andrew,* in which Christ sees the two men in a boat casting their nets in the Sea of Galilee. They join him, becoming the first of the apostles. The initials *D* and *Q* show, respectively, Saint Andrew holding the cross on which he was crucified and Saint Peter holding the key to heaven.

Bolognese illumination blossomed during the thirteenth and fourteenth centuries, due in part to the rise of the book trade in the university town. This book's anonymous illuminator, the Master of the Brussels Initials, was a student of Niccolò da Bologna (circa 1330–1403/4), one of the finest Italian illuminators of the fourteenth century. The strong local colors, the intensity of the holy men's gazes, and their bulky robes probably reflect the influence of Niccolò. On the other hand the border teeming with drolleries, beasts, and acanthus leaves is this master's own innovation. Within a decade of painting the missal the Master of the Brussels Initials moved to Paris, where he became a major figure in the French International style of manuscript illumination. His distinctive style of decorative borders was widely imitated there.

Cardinal Cosimo de' Migliorati (circa 1336–1406) commissioned this book before his election to the papacy in 1404 as Innocent VII. His arms in the lower margin are overpainted with the papal tiara and arms of the Antipope John XXIII (d. 1419), who was elected pope in 1410 and deposed in 1415. Both were pope during the Great Schism of the West (1378–1417), when a second pope resided simultaneously in Avignon.

TK

Incipit proprium
scorum de missali.
In uigilia sci
Andree apli. Introitus:-
Dominus
secus ma
re galilee
uidit du
os fratres
petrum
et ãdreã
et uocauit
eos. uenite post me faciam uos fie
ri piscatores hominũ. Ps. At
illi relictis retibus et naui sequuti
sũt dnm. V. Gloria. Orõ.
Quesu
mus
oips
deus:
ut be
atus
ãdre
as apls tuum pro nobis im
ploret auxilium: ut a nostris
reatibus absoluti. a cunctis etiã
periclis eruam. P. Pro sco
saturnino mre. oratõ.
Deus qui nos bti satu
rnini mris tui cõcedis natali
cio perfrui: eius nos tribue
meritis adiuuari. P. Epl'a. re
quire in uigilia unius apostli.
Benedictio dni sup capud
iusti. Gr. Nimis honorati sũt
amici tui ds. nimis cõfortatus est
pricipatus eorum. V. Dinu
merabo eos et sup arenã multiplica
buntur. Seqn. sci eu. s'm iohem.
In illo tpr. Stabat iohes.
et ex discipulis ei' duo.
Et respiciẽs ihm ambulantẽ
dic. Ecce agnus dei. Et audi
erunt eum duo discipli loq'n
tem: et sequuti sunt ihm. Cõ
uersus autẽ ihs et uidẽs eos

Incipiunt hore de domina nr̄a

Domi
ne la
bia
mea
ape
ries
Et
os
meum annunciabit laude
tuam.

Deus in adiutorium me
um intende. domine ad ad

26 Book of Hours
Probably Utrecht,
circa 1405–1410

210 leaves, 16.4 x 11.7 cm
(6 1/2 x 4 5/8 in.)
Ms. 40; 90.ML.139

Plates: Masters of Dirc van Delf,
Initial *D* with *Madonna and Child,*
fol. 14
The Entombment, fol. 79v

See pages 64–65

The transition from the fourteenth to the fifteenth century saw a continuous flowering of manuscript illumination in the farthest corners of Europe. One of the newest centers was the northern Netherlands (modern Holland), where a style of court illumination blossomed under the benevolent patronage of Albrecht of Bavaria, Count of Holland (r. 1389–1404). Gathering artists, musicians, and intellectuals at his court in The Hague, Albrecht engaged the eminent Dominican theologian Dirc van Delf as court chaplain while commissioning illuminated copies of Dirc's writings. Called, suitably, the Masters of Dirc van Delf by scholars, several of these anonymous artists illuminated this book of hours. The Masters of Dirc van Delf formed one of the first important illuminators' workshops of fifteenth-century Holland. Because of the workshop's association with the court chaplain and the origins of its art in painting at Albrecht's court, it seems likely that a member of Albrecht's family or one of his courtiers commissioned the Getty manuscript.

The Hours of the Virgin open with an image of the Virgin and Child. She is shown crowned as Queen of Heaven, but seated on the ground, a reflection of her humility and thereby an example for the reader. Her sweet, youthful face, the full modeling of the amply robed body, and the soft lighting are characteristics of painting and illumination in Northern Europe at this time. (Compare, for example, no. 28, painted not far away in Cologne.)

Books of hours not only fostered devotion to the Virgin Mary but also provided meditations on the meaning of the story of Christ. As this manuscript illustrates, their miniatures complement the texts, engaging the emotions and fostering empathy with Jesus for his supreme sacrifice. In *The Entombment,* the Virgin, Joseph of Arimathea, and Nicodemus, their eyes filled with sadness, gently lay Christ's body into the tomb. The Virgin contemplates her son's face and with it the meaning of his death, just as the viewer is invited to use this image to meditate on the same truths. The artists show the caretakers' broad forms extending beyond the confines of the painted frame; by implication they move closer to the viewer's experience. TK

27 Rudolf von Ems, *Weltchronik*
Regensburg, circa 1400–1410

309 leaves, 33.5 x 23.5 cm
(13 3/16 x 9 1/4 in.)
Ms. 33; 88.MP.70

Plates: *The Construction of the Tower of Babel,* fol. 13
The Israelites' Fear of the Giants and the Israelites Stoning the Spies, fol. 98v

See pages 68–69

Rudolf von Ems, a German knight and a prolific writer, composed his *World Chronicle* toward the middle of the thirteenth century. Left uncompleted at his death around 1255, the *Weltchronik* sought to trace history from Creation to the present. The chronicle depended to a large extent on the events of the Bible for its narrative, as is evident from its division into six ages—Adam, Noah, Abraham, Moses, David, and Christ. Rudolf's text, which comprises some thirty-three thousand lines of rhymed German verse, ends in the middle of the story of King Solomon.

Rudolf moved away from the courtly romances and lyrics that characterized Middle High German literature and returned to the tradition of writing more sober history. Interwoven with his biblical narrative is information relating to the Trojan War and Alexander the Great, to name but two examples. Rudolf's *Weltchronik* enjoyed an unusual popularity and was itself used as a model for later vernacular chronicles.

This early fifteenth-century manuscript is one of numerous illustrated copies of the *Weltchronik* and contains other historical texts, including a *Life of the Virgin Mary.* Of the volume's almost 400 miniatures, 245 illustrate incidents in Rudolf's work. *The Construction of the Tower of Babel,* showing King Nimrod at left supervising the operation, depicts a variety of building procedures that probably mirror medieval practices closely. In an episode from the Book of Numbers (chapters 13–14), the Israelites react to the news brought by the twelve spies that the land of Canaan is inhabited by giants, represented here as contemporary knights. While some engage in animated debate, others seek to stone Joshua and Caleb, the two spies who voiced their faith in God's providence. The art in this German *Weltchronik* is characterized not only by vivid coloring and bold brushwork but also by the agitated movement and the psychological intensity of the figures. These features stand in contrast to the jewel-like color, courtly dress, and demure physical types of the International style that flourished in European painting and manuscript illumination at the beginning of the fifteenth century (nos. 25–26, 28–32). ASC

26

In disen sellen stunden
Des werches si begunden
Und heten in der tage zil
Des werches gächs also vil
Gemachet daz ez sich gezoch
Mer dann funf tausent schrit hoch
Und sibenzig vn̄ nawn hundert
Und vier schrit auz gesundert
Mit zwain vn̄ sibenzig ekken waz
Der selb turn alz ich ez laz

Si heten funden ein lant
Daz nieman waz bechant
Ein pezzer lant anderswa
Läwt vnd gutes waz alda
Mit reichait mer danne vil
Vnd vest nach dez wunsches zil
Gepawen vestichleich
Daz lant waz gut vnd reich
Auch heten si daz inn
Geseheen ze vngewinn
Daz geslacht von enach
Do erschrakt daz her vnd sprach
O we der vorchtleichen not
O we warn wir dann tot
In egypten gelegen
Seint wir vns nw muzē bewegē
Daz wir ligen tot von disen
Laidigen vngehewrn risen
Daz waz vns pez ergangen

Dann daz vns nw gevangen
Werdent dort weib vnde chint
Dy von vns gepoarn sint
Vnd chiesen einen hauptman
Der vns fur wider dan
Caleph vnd Iosue
Wen tet ir zweifel red we
Vnd also we daz si zehant
Av in zarten daz gewant
Von der zweifels märn
Den tumben zweifelärn
Straften si ir mär do
Vnd sprachen zw in also
Seit nicht widersprüchig got
Daz ir an seinem gepot
Icht werdet zweifelhaft
Wir haben so vest chraft
Gein den lawten dy dort sint
Daz wir si ezzen als ein rint
Daz gras auf ainer waid tut

28 Two Miniatures, Perhaps from a Manuscript
Cologne, circa 1400–1410

23.6 x 12.5 cm (9 5/16 x 4 7/8 in.)
Ms. Ludwig Folia 2; 83.MS.49

Plate: Master of Saint Veronica, *Saint Anthony Abbot Blessing the Animals, the Poor, and the Sick,* leaf 2

Cologne on the lower Rhine River was a major artistic center throughout the Middle Ages; led by a painter called the Master of Saint Veronica (fl. circa 1390–1410), the town produced several of the finest painters of the fifteenth century. Cologne's proximity to Dutch and Flemish towns placed it within a nexus of burgeoning artistic creativity, especially in painting and manuscript illumination.

The Master of Saint Veronica shows the fourth-century hermit saint Anthony Abbot blessing the sick, the poor, and animals. He stands on a pedestal dressed in the black mantle with *Tau* sign (*T*) and white robe of the Order of the Hospitallers of Saint Anthony, and in stylish and costly shoes. He holds the crosier of an abbot. The pedestal resembles the socles that support polychrome devotional carvings of saints of the time, a reminder to the viewer that this is not merely a narrative scene. The saint himself is an object for our veneration. The Hospitallers of Saint Anthony dedicated their ministry to caring for the sick and infirm. One of the most widely venerated of saints during the Middle Ages, the hermit Anthony was invoked for assistance against various diseases, especially the one popularly known as Saint Anthony's fire (*erysipelas*). A particularly widespread and virulent malady of the Middle Ages, erysipelas caused gut-wrenching pain, contortions, and hallucinations. Its consequences included amputation of limbs and inevitable death.

Cologne had an important church dedicated to Saint Anthony with a hospital run by the order. It was rebuilt during the 1380s, less than a generation before the Museum's miniatures were painted. By some accounts, the Abbot of Saint Anthony's in Cologne blessed the animals on the saint's feast day each year (January 17). It seems likely, therefore, that the Master of Saint Veronica painted this miniature and its companion expressly for a book or small altarpiece for that church or a chapel in the adjoining hospital.

The brilliant colors, sweet and tender facial expressions, courtly and elegant costumes, and nuanced modeling reflect a style of painting that links such diverse centers as Cologne, Utrecht, Paris, Prague, and London around 1400. Scholars call this phenomenon the International style.

TK

29 Missal
from the Collegium Ducale
Vienna, circa 1420–1430

307 leaves, 41.9 x 31 cm
(16½ x 12³⁄₁₆ in.)
Ms. Ludwig V 6; 83.MG.81

Plate: *The Crucifixion,* fol. 147v

The International style takes its name from the art that was created in such disparate European centers as Paris, Utrecht, Cologne, and Prague. The broad stylistic unity within architecture, sculpture, painting, and manuscript illumination was the result in part of the increased movement of artists who were attracted by courts with extensive ties throughout the continent. In Eastern Europe, Bohemian Prague became the leading political and cultural hub as capital of the Holy Roman Empire under Charles IV, himself raised and educated in France.

Another, less familiar, center of art production was Vienna, where this missal was made. Its painters, including an illuminator known only as Michael, have been identified from their work in other commissions from Vienna, Bohemia, and Slovakia. The missal is thus a witness to the cross-fertilization in Central European art at this time. The association of artists probably trained in Bohemia, but working together in Vienna indicates the city's increasing importance. That Viennese patrons enlisted such artists suggests, too, that they may have been seeking to compete with the powerful Bohemian court.

In this image of the Crucifixion, Jesus' drooping head, emaciated torso, and frail arms show his suffering on the cross. At the bottom of the page Jesus appears as the Man of Sorrows, the risen Christ who displays his wounds to the devout for contemplation. The juxtaposition of the two images demonstrates that resurrection and salvation are already inherent in the act of Crucifixion. The miniature simultaneously conveys a sense of refined elegance typical of the International style. This is evident first in the subdued coloring, as the delicately ornamented pink background highlights the primary blues and greens. The gentle sway of the figures and the sinuous contours of their robes are characteristic of this stage of late Gothic painting.

According to a treasury inventory written in the manuscript in 1508, the book then belonged to the Collegium Ducale. Established in 1384, this theological faculty was part of the Vienna University, which had been founded in 1365 by Duke Rudolf IV of Austria. It cannot be said whether or not the missal was originally made for the Ducal College.

ASC

30 Giovanni Boccaccio,
Des Cas des nobles hommes et femmes
Paris, circa 1415

318 leaves, 42.5 x 29.3 cm (16¾ x 11 9/16 in.)
Ms. 63; 96.MR.17

Plate: Boucicaut Master and Workshop, *The Story of Adam and Eve,* fol. 3

The Florentine poet and man of letters Giovanni Boccaccio (1313–1375) is one of the fathers of Renaissance humanism. Within a generation of his death, Boccaccio's writings were already popular outside of Italy. A number of them, including *The Decameron*—the one most read today—were translated into French under the patronage of such august figures as Philip the Bold, Duke of Burgundy (1342–1404) and John, Duke of Berry (1340–1416). In France at that time the most beloved by far was this text of *The Fates of Illustrious Men and Women.* It relates the stories of notables from biblical, classical, and medieval history. Laurent de Premierfait (d. 1418), who translated Boccaccio's works, embellished the original with many colorful tales from other authors including the ancient Romans Livy (59 B.C.–A.D. 17) and Valerius Maximus (circa 49 B.C.–circa A.D. 30).

Boccaccio begins the book with the lives of Adam and Eve, since their sin gave rise to the calamities that would befall humankind. Ingeniously organizing the sequence of events around the tall hexagonal walls of the Garden of Eden, the Boucicaut Master shows us the Temptation of Adam and Eve in the center. The first couple are driven from the garden through a portal at the left, and beyond the garden walls they assume their fates toiling in the fields and spinning. In the foreground right we see Adam and Eve, now elderly and stooped, approaching the author to tell their story. Boccaccio is elegantly robed in red. The artist has created an elaborate frame that encloses both the miniature and the opening lines of the text. It includes a sequence of painted vignettes depicting the Creation of the World, commencing at the upper right and proceeding clockwise.

The first quarter of the fifteenth century proved to be one of the most original and influential epochs of Parisian manuscript illumination, due in significant part to the genius and industriousness of the Boucicaut Master. With the aid of numerous highly trained collaborators, this artist's innovative work became known throughout Europe and influenced not only the direction of French illumination for more than a generation but that of Flemish painting as well. TK

De adam et eve premier chappitre com
mencant en latin Maiorum nostrorum etc.

uant je conside
re et pense en di
vers les manieres
les plourables ma
leuretes de nos pre
decesseurs/ a celle
fin que du grant
nombre de ceulx
qui par fortune ont este trebuschiez je:
preinsisse au commencement de ce livre
aucun prince terrien asses digne destre
premier entre les maleureux/ Et veci

deux viellars se adresserent devant moy
si tres aagiez et si anciens quil sembloit
que ilz ne peussent trayner leurs mem
bres tremblans. Lun de ces deux
viellars cest assavoir adam me raison
na. Et dist beau nepveu Jehan boccace
qui cerches et enquiers lequel tu met
tes premier au reng des maleureux:
Je vueil que tu saiches comme vray est
que ainsi comme nous deux qui som
mes les premiers homme et femme:
fais a limaige de dieu qui par le moyen
et admonnestement de lui avons premiers
a mieu et emply les sieges de paradis par

31 Book of Hours
Paris, circa 1415–1420

281 leaves, 20.4 x 14.9 cm
(8 1/16 x 5 13/16 in.)
Ms. 22; 86.ML.571

Plate: Boucicaut Master, *All Saints*, fol. 257

Toward the end of the fourteenth century Eustache Deschamps (circa 1346–1406), a poet and artist at the court of Charles VI of France, mocked the widespread demand for illuminated books of hours among middle-class women. He held up the fashion as a display of vanity and shallow materialism:

> A book of hours, too, must be mine
> Just as a nobleman desires
> Let it be splendidly crafted in gold and azure
> Luxurious and elegant . . .

To judge from the books that survive, Deschamps's complaint had no impact. The demand for richly decorated books of hours exploded at the beginning of the fifteenth century, and Paris experienced one of its greatest flowerings as a center of manuscript illumination. Aided by assistants and collaborators, the anonymous artist called the Boucicaut Master, the city's finest illuminator, supplied the market generously with books of hours. This one, created for a rich bourgeois woman named Margaret, shows the expensive pigments the Boucicaut Master employed to dazzle his clients and the very high level of artistic refinement he achieved.

In the page shown here, a suffrage (or prayer invoking the intercession of saints) for All Saints is illustrated by the holy men and women robed in elegant and rich colors of rose, burgundy, gold, orange, and several shades of blue. A stock (even dull) subject, the artist enlivens it through the alert and engaged expression of each of the figures. The lifetime of the Boucicaut Master (fl. 1400–1420) saw the dawn of a tradition in Northern European painting that imparted fresh attention to the interior lives of its subjects. This interest in characterization and human psychology has remained an essential element in European painting since that time. TK

Succurrite michi ho
die et omni tempo
re omnes sancti

31
In illo tpe. Apprehen
dit pylatus ihesum
et flagellauit eum

Deus in adiutori
um meum intende.
Domine ad ad

32 Book of Hours
Probably Paris, circa 1415–1425

247 leaves, 20.1 x 15 cm
(7 15/16 x 5 7/8 in.)
Ms. 57; 94.ML.26

Plates: Spitz Master,
The Road to Calvary, fol. 31
The Flight into Egypt, fol. 103v

See pages 78–79

This manuscript was produced in the orbit of the Limbourg Brothers, who painted only a handful of manuscripts and worked primarily at the court of John, Duke of Berry (1340–1416). The books they painted for him are among the greatest of the later Middle Ages. Some of the Getty book's miniatures, including that of *The Road to Calvary* (fol. 31), are adapted from illuminations by the Limbourgs. Here Christ is shown barefoot but in a fine robe trimmed in gold thread, carrying the cross through the city gate of Jerusalem toward Calvary. A pair of soldiers pull and push him along his path. In the distance, the remorseful Judas is shown having hanged himself. To heighten the page's spiritual and contemplative character, the illuminator has added to the border angels carrying the Instruments of the Passion: a crown of thorns, a spear with a sponge, utensils for human flagellation, pliers for removing nails from the cross, and the nails themselves.

In the painting of the garments, the use of expensive materials (including burnished silver), and the tender expressions of the figures, this miniature epitomizes the refinement and elegance of court art at the beginning of the fifteenth century. When not copying, the Spitz Master shows a different side of his personality. In *The Flight into Egypt,* Joseph leads Mary on a donkey to escape the cruel Herod, King of Judea, who has decreed the death of all young children in an effort to destroy the newborn Christ Child. The illuminator shows the Holy Family journeying through a hilly, seemingly enchanted landscape. At the left, Herod's men are shown in pursuit; their enormous heads peeking over the horizon dwarf their surroundings. The exaggerated scale of the soldiers and buildings contributes to the sense of danger and enchantment in the Holy Family's escape.

The border illustrates the Miracle of the Wheat Field, another incident on their flight. When the Family passes a worker sowing wheat in the fields, the Virgin asks him to inform their pursuers that he saw the Family while sowing. The illuminator depicts the soldiers' arrival shortly thereafter, when the wheat has miraculously grown tall, so the sower's true account suggests to the soldiers that they are well behind. TK

33 Hours of Simon de Varie
Tours and perhaps Paris, 1455

97 leaves, 11.5 x 8.2 cm
(4½ x 3¼ in.)
Ms. 7; 85.ML.27

Plate: Jean Fouquet, *Simon de Varie in Prayer Before the Virgin and Child,* fols. 1v–2

Portraiture was one of the great achievements of fifteenth-century painting, especially in Northern Europe. The French artist Fouquet (d. 1478/81) was one of its most accomplished practitioners. In his youth he portrayed Pope Eugene IV (1431–1447), and he subsequently received many commissions from the court of the French king Charles VII (r. 1422–1461).

Among the court officials was Simon de Varie who, recently ennobled, had risen to a post in the royal treasury under Charles VII. He is shown here as a handsome youth kneeling in prayer before the Madonna and Child. While earlier books of hours had featured similar devotional portraits, what is unusual in the Varie Hours are the patron's arms, also painted by Fouquet on the backs of the two leaves. This ensemble of four illuminations, probably placed at the front of the book, offers an eloquent statement of the patron's pride in his lofty new status in society.

Although Varie was not a soldier, he wears a suit of armor and a surcoat with his personal heraldry. Behind him a female attendant supports a Varie escutcheon crowned with helmet and crest. Similar coats of arms (all now partially overpainted) and Simon's personal mottoes appear in the borders. The motto in the upper border, *Vie à mon désir* (Life according to one's desire), is an anagram of his name.

The complete Varie Hours includes forty-nine large miniatures by four artists and several dozen other vignettes and historiated initials. A later owner divided the book into three volumes. The Getty Museum owns one and the two others belong to the Royal Library in The Hague in the Netherlands. TK

34 Book of Hours
Tours, circa 1480–1485

145 leaves, 16.3 x 11.6 cm
(6 7/16 x 4 9/16 in.)
Ms. 6; 84.ML.746

Plate: Jean Bourdichon,
The Coronation of the Virgin, fol. 72

This ceremonious, joyful miniature shows two angels crowning the Virgin Mary as Queen of Heaven. God the Father offers his blessing from heaven as he displays an orb, the symbol of his universal dominion. Below, an assembly of angels bears witness to the hallowed event. Painted by Jean Bourdichon of Tours (circa 1457–1521), official painter to four successive French kings, this manuscript contains some of his earliest known work. Bourdichon succeeded Jean Fouquet as royal painter, and his art shows the powerful impact that Fouquet's innovations exercised on French illumination in the second half of the fifteenth century.

While Bourdichon probably had not visited Italy at this early moment in his career, he learned from Fouquet principles of Italian Renaissance painting. These include the use of symmetry and geometric form to compose the miniature; for example, he arranges the angels at the feet of the Virgin in an ellipse. Bourdichon probably also learned from Fouquet to paint both spiritual and physical light. Golden rays of divine light emanate from the Virgin (against a celestial curtain of dense blue), while the same light softly models the draperies of the two angels and the faces of those below. One of the more subtle effects is the slight twist in the axis of the crowning angels, which relieves the composition's marked symmetric character and strengthens the illusion of recession.

The initials *I* (or *J*) and *K* appear four times in the border, the *I* embraced by a loop that forms the arms of the *K*. Such letters usually are the initials of a husband and wife who commissioned the book. The prominence of several prayers to Saint Catherine of Alexandria suggests that the *K* may refer to an owner named Catherine.

TK

Ad complectorium
Onuerte nos deus
salutaris nr
Et auerte iram
tuam a nobis

Comment langele apres
les tourmens des larrons
et robeurs mena lame
du chãllier aux peines
des gloutons et des forni
cateurs. Et comment ilz
y veirent executer en tres
cruelz tourmens. Le c.

Ainsi comme
langele avoit et
lame du chãllier
passe les tourmens des
larrons et robeurs ilz se
misrent a la voye. Et aussi
comme ilz sen aloient par
tenebres veirent une tres
grant mansion ouverte
Icelle mansion certes est
oit aussi grande comme une
bien haulte montaigne
Et estoit toute ronde com
me ung four. Et partout
de icelle mansion une telle
flamme que par mille pas
environ delle tout ce quelle
attaindoit brulloit et ar
doit. Et lame du chãllier
quy avoit desia esprouve
semblables tourmens
aler ne vouloit avant si
dist al angele quy le con
duisoit. Las chier sire

35 *Les Visions du chevalier Tondal*
Ghent and Valenciennes, 1475

45 leaves, 36.3 x 26.2 cm
(14⁵⁄₁₆ x 10⁵⁄₁₆ in.)
Ms. 30; 87.MN.141

Plates: Attributed to Simon Marmion,
The House of Phristinus, fol. 21v
Tondal Appears to Be Dead, fol. 11
The Joy of the Faithfully Married,
fol. 37

Visions of a journey through hell constitute one of the most popular medieval literary genres. Before the time of Dante, *The Visions of Tondal,* the story of a morally errant Irish knight whose soul embarks on such a journey, was the most widely disseminated. Written by Marcus, an Irish monk, in Regensburg (Germany), around 1149, its Latin text came to be translated into fifteen different languages over the next three hundred years. This French translation, dated March 1475, was undertaken for Margaret of York, the Duchess of Burgundy and consort of Charles the Bold. Their initials appear in the lower border. The duchess's copy captures in twenty scenes much of the narrative's vivid, often terrifying detail. Briefly, the story of the young and selfish Tondal unfolds as follows. While visiting a friend to collect a debt, he collapses and seems to be dead. In this state an angel leads his soul on a journey, protecting it along the route from the demons and torments of hell. Tondal's soul witnesses the terrible punishments meted out for various sins, such as the cavernous House of Phristinus where gluttons and fornicators are tormented by flames and infernal monsters. The soul then passes to purgatory in a journey toward paradise. Along the way it encounters those who have lived better lives and enjoy the prospect of redemption. At the end Tondal realizes the errors of his ways and returns to a life of Christian penitence.

Simon Marmion (circa 1420–1489), a favorite painter and illuminator of the Burgundian court, appears to have created the miniatures in this volume. Departing from his characteristic use of pastel tones (which appear, for example, in *The Joy of the Faithfully Married*), Marmion conjures up the murky darkness and flickering brightness of hell, all vaporous and fiery, along with its monstrous inhabitants. TK

36 Book of Hours
Provence, circa 1480–1490

198 leaves, 11.5 x 8.6 cm
(4½ x 3⅜ in.)
Ms. 48; 93.ML.6

Plates: *The Visitation*, fol. 34
Georges Trubert, *Sorrowful Madonna*, fol. 159

The main illuminator of this book is Georges Trubert (fl. circa 1469–1508), who served King René I of Anjou (1409–1480) at his court in Provence for the final decade of the ruler's life. He then remained in southern France for another ten years. Himself a poet and writer, René was also a visionary patron of the arts. The book contains several miniatures that allude to specific paintings he owned, among them an older painting or icon representing a weeping, or sorrowful, Madonna that is copied in an apparently imaginary altar-reliquary. Another artist in the book, who painted *The Visitation*, shows visual similarities to both Trubert's miniatures and those of artists active to the north in the Loire valley.

The illuminators of this book explore diverse ways to make painted objects appear palpable and three-dimensional against the flatness of the page. The border of the miniature showing the meeting of the Virgin Mary, now pregnant with Jesus, and the elderly Elizabeth, who bears the future John the Baptist in her womb, depicts birds, foliage, and music-making drolleries. It is painted in brown monochrome. This gives it the character of a wood carving in shallow relief. The leaves of this "carving" curl off the edges of the painted border onto the real page, heightening the impression of depth.

More unusual and enigmatic is the miniature of an altar-reliquary in metalwork containing the *Sorrowful Madonna*. The lustrous altar is shown with wings of gold, silver, and enamel opened up, the right wing appearing to cast a shadow on the page. The altar sits on a grassy clod of earth that is in turn supported by two bronze figurines of lions. Along with the vines of columbine emerging above the shrine, this curious devotional object also casts a shadow. A piece of parchment with the words *O Intemerata* (O Immaculate Virgin) is painted to appear tacked below the Madonna, the lower right corner of the parchment curling free of its tack. *O Intemerata* are the opening words of a prayer to the Virgin that continues when the reader turns the page. TK

O INTEMERATA

Incipiūt septē
psi· penitēnales
a Ne reminiscar
Domine ne ī fu

37 Book of Hours
Probably Ghent,
circa 1450–1455

286 leaves, 19.4 x 14 cm
(7 5/8 x 5 1/2 in.)
Ms. 2; 84.ML.67

Plates: Master of Guillebert de Mets, *The Last Judgment* and *King David in Prayer,* fols. 127v–128

See pages 88–89

Although a book of hours contains a common core of devotional texts, an ambitious version such as this one would have numerous supplemental texts and variations. Similarly, an illuminated book of hours could have a complex and extensive program of painted decoration, and the most far-reaching were often vehicles for artistic innovation. While borders are usually subordinate to miniatures, this manuscript reverses the relationship to a degree. The animated, monumental foliage of its borders captures the attention, and, as this two-page opening shows, the borders give unity to the whole spread.

In the borders shown here supple lilies flourish, their petals rhythmically curling and swelling, as if the flowers were opening on the pages as we turn to them. The petals pass over and under the thin frames of the miniatures, making the border's presence palpable in a way the miniatures are not. Banderoles—fluttering ribbons inscribed with text—weave paths through the borders, into and out of *The Last Judgment* on the left and *King David in Prayer* on the right. Banderoles emanate from horns blown by angels of the Last Judgment and pass under the frame at the top of the David miniature. The assembly of naked souls gathered in the opening of a lily in the left border contributes further to this integration. They are the dead resurrected to face the Last Judgment.

The two miniatures mark the beginning of the Seven Penitential Psalms, a major text in a book of hours. These psalms are meditations on human frailty and petitions to the Lord for mercy, succor, and salvation, serving to prepare the soul for the Last Judgment. The first of the seven is Psalm 6, commencing with the line: "O Lord, do not condemn me in your anger . . . " (*Domine ne in furore tuo arguas me* . . .). King David is shown penitent, his harp at his side.

The Flemish illuminator called the Master of Guillebert de Mets (fl. 1420–1450) illustrated this opening and a number of other major decorations in the book. Trained in Paris or by Parisian illuminators who worked in Flanders, he lived in or near Ghent toward the end of his life when the book was made. TK

38 Prayer Book of Charles the Bold
Ghent and Antwerp, 1469

159 leaves, 12.4 x 9.2 cm
(4 7/8 x 3 5/8 in.)
Ms. 37; 89.ML.35

Plates: Lieven van Lathem, *Christ Appearing to Saint James the Greater,* fol. 22
Text Page, fol. 30v
Lieven van Lathem, *All Saints,* fol. 43
Attributed to the Master of Mary of Burgundy, *The Deposition,* fol. 111v

See pages 92–93

The household accounts of the Burgundian dukes record payments in 1469 to the scribe, the illuminator, and the goldsmith (who fashioned the clasps for the binding) of this elegant and costly prayer book. The duke himself, Charles the Bold (1433–1477), the son of the bibliophile Philip the Good, commissioned it. The duke paid Lieven van Lathem (circa 1430–1493) of Antwerp and Nicolas Spierinc of Ghent (fl. 1455–1499) for their illumination and writing, respectively, of this book. The original binding of this manuscript was replaced by the early sixteenth century. The work of the goldsmith Ernoul de Duvel is lost.

The diminutive volume is distinctive for the embellishment of each page, not only the illuminated pages but also those without any painted decoration. Spierinc, one of the most original of scribes, filled the borders of text pages with exuberant *cadelles,* whose lush decorative quality complements the illuminated pages. On the text page reproduced here, delicate painted drolleries further enliven the margin.

The miniatures, measuring only around three by two inches, are meticulously detailed, often with atmospheric landscapes that seem to extend for miles. Indeed, while Antwerp became famous as a center of landscape painting only in the sixteenth century, its citizen van Lathem paved the way in such miniatures as *Christ Appearing to Saint James the Greater*. The lazily winding river pulls the eye to a distant horizon. The borders are every bit as compelling as the miniatures, with their grotesques and playful figures that descend from the tradition of marginal decoration in Gothic manuscripts. Among the men and monsters gamboling in the dense foliage of the monochrome border of the same page, a lion has pinned a nervous soldier to the ground.

Although van Lathem painted most of the book's thirty-nine original miniatures, several collaborators enabled him to complete the illumination. The most gifted was the painter of the moving *Deposition,* which anticipates in its depth of feeling and the nuanced rendering of the fragile corpse of the dead Christ the mature art of the Master of Mary of Burgundy, the doyen of Burgundian illuminators (see no. 42). It is perhaps one of his earliest works. Here the border vignette of Adam and Eve mourning the death of Abel offers an Old Testament prefiguration of the mourning over the body of Christ as he is taken down from the cross. TK

Corpora scõrum
in pace sepulta

Deus in ad
iutorium

39 Fifteen Leaves from David Aubert, *Histoire de Charles Martel*
Brussels and Bruges, 1463–1465 and 1467–1472

22.6 x 18.4 cm (8 15/16 x 7 1/4 in.)
Ms. Ludwig XIII 6; 83.MP.149

Plate: Loyset Liédet, *Gerard and Bertha Find Food and Sustenance at a Hermitage,* leaf 5

Philip the Good, Duke of Burgundy (1419–1467), not only expanded dramatically the size of the duchy of Burgundy but also built one of the great libraries of the fifteenth century. It contained more than seven hundred volumes. His vast patronage fostered the most important era of manuscript illumination in Flanders, one that continued long after his death. This miniature and fourteen others at the Getty were once part of the *Histoire de Charles Martel* (History of Charles Martel) that was written for him in four volumes—for a total of two thousand leaves or four thousand pages—by court scribe David Aubert over a period of several years (1463–1465). Philip traced his ancestry to Martel (r. 714–741), the grandfather of Charlemagne, an outstanding military leader and the ruler of the Frankish kingdom (which encompassed modern France and Germany). Late medieval knights undoubtedly enjoyed reading the adventures of such ancient heroes, and Philip would have drawn inspiration from his exploits.

Several years after Philip's death, the illumination of this extravagant undertaking had barely begun. In 1468 ducal accounts show payments to one Pol Fruit of Bruges for painting the initials in the third volume. A year or so later Philip's son and heir, Duke Charles the Bold, hired Loyset Liédet to paint the book's 123 miniatures. During the 1460s and 1470s, the prolific Liédet worked in Hesdin in northern France and in Bruges. He received payment for miniatures in this book in 1472. In total the manuscript took a decade to produce.

The illustration shown here represents Gerard de Roussillon, the great hero of the Burgundians and a rival of Charles Martel, with Bertha, his wife. After being robbed of their horses, they are offered food and find drink at a spring.

The four volumes of the book, still preserving 101 of the original miniatures, belong to the Bibliothèque Royale in Brussels, which acquired the core of Philip the Good's library. TK

courrouchiee que plus elle nen pouoit / Et moult souuent regrettoit sa suer / et tous les iours ne attendoit aultre chose fors que elle retournast vers elle a reffuge / Mais a tant sen taist vng petit listoire / Et retourne a parler du mal fortune prince monseigneur gerard de roncillon

Comment le noble prince gerard de roncillon deuint charbonnier par fine contrainte de pourete et misere.

L'Ancienne histoire racompte que quant le noble prince gerard de roncillon se fut party des marchans de france lesquelz luy auoient dit nouuelles de la mort du roy othon de hongherie / et des francoys qui tantost

40 Quintus Curtius Rufus,
Livre des fais d'Alexandre le grant
Lille and Bruges,
circa 1468–1475

237 leaves, 43.2 x 33 cm (17 x 13 in.)
Ms. Ludwig XV 8; 83.MR.178

Plate: Attributed to the Master of the *Jardin de vertueuse consolation, Alexander and the Niece of Artaxerxes III,* fol. 123

Alexander the Great (356–323 B.C.), King of Macedonia, conquered much of the ancient world. He gained vast territories extending from the eastern Mediterranean to northern India. His fame endured throughout the Middle Ages and his name still evokes wonder today. The emergence of humanism in Northern Europe during the second half of the fifteenth century fostered the desire for a reliable account of his exploits, one no longer encumbered with the stuff of legend and romance that had accrued during the Middle Ages. Vasco da Lucena, a Portuguese diplomat and humanist at the Burgundian court, chose the text of the ancient Roman historian Quintus Curtius Rufus, who appears to have lived in the first century, as the most reliable of the ancient accounts. Vasco endeavored to translate it into French while replacing portions that were lost. His effort, dedicated to Charles the Bold, Duke of Burgundy, enjoyed popularity at the court and throughout Flanders and France. The Getty copy was probably made for a nobleman in the circle of the duke.

In the miniature illustrated here the niece of the Persian king Artaxerxes III (r. 358–338 B.C.) is shown kneeling before Alexander. The conqueror had noticed her among his Persian prisoners. Because she is a member of a royal family, he decides to free her and return her belongings. Vasco detailed such incidents to provide a balanced picture of his subject's character; elsewhere in the text he shows us Alexander's cruelty, vanity, and other frailties. The book's anonymous painter also illuminated other large-format volumes for Burgundian noblemen. His art shows affinities with that of the Antwerp illuminator Lieven van Lathem (no. 38). Jean du Quesne, who transcribed this copy of Vasco's text, was himself the translator of other humanist texts.

Large histories such as these were read aloud to their owners from a lectern. Alexander's exploits must have appealed especially to the knights of the Burgundian court, while the convention of depicting ancient personalities in the contemporary dress of the court gave the stories particular immediacy. The fourteen miniatures of the Getty *Alexander* are colorful and filled with action. They show battles and conquests, assassinations and court intrigue.

TK

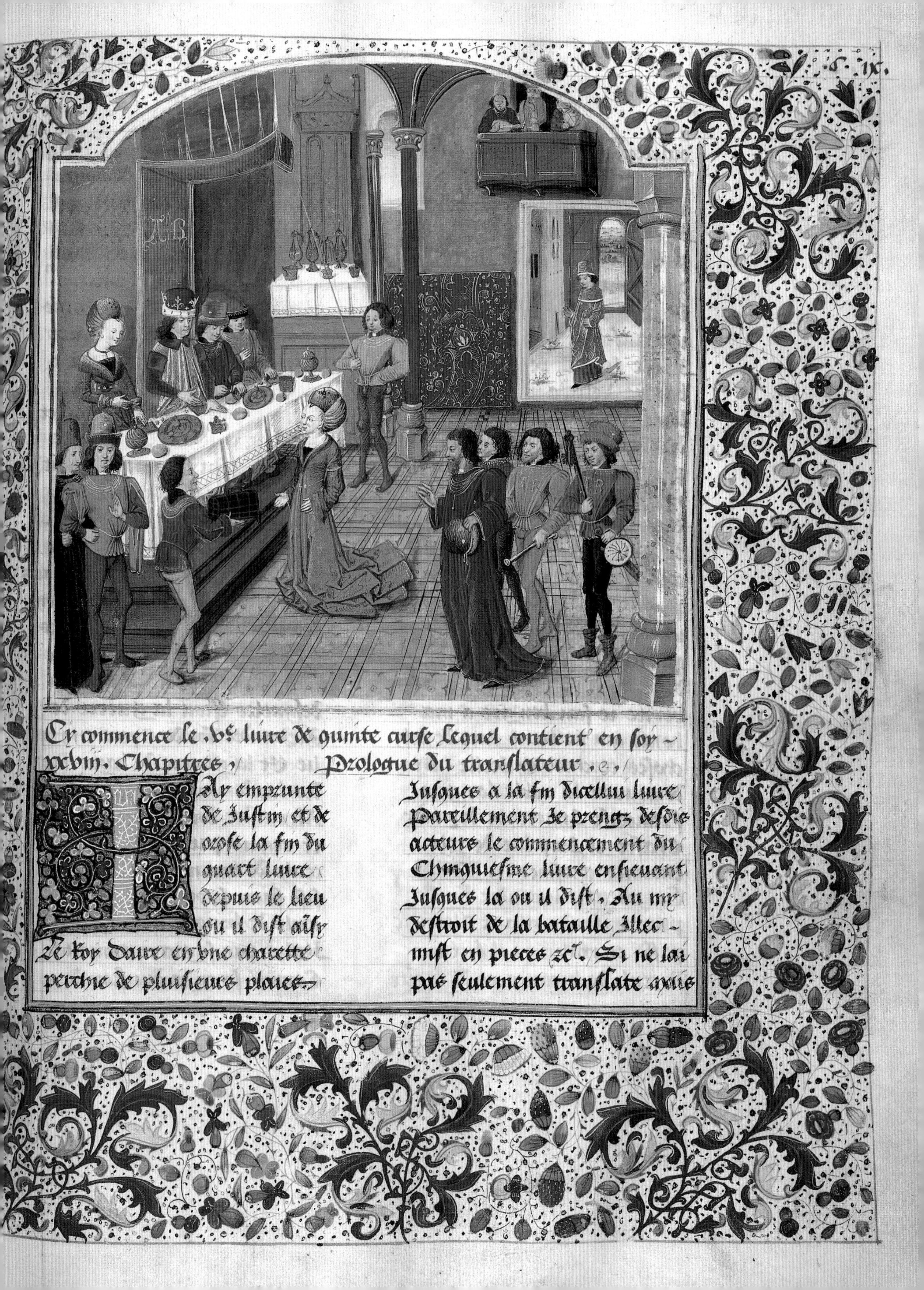

Cy commence le .v^e. livre de quinte curse lequel contient en soy-
xxviii. Chapitres. / Prologue du translateur.

JAy emprunte de Justin et de orose la fin du quart livre depuis le lieu ou il dist ainsy Le roy daire en une charette perchie de plusieurs plaies. Jusques a la fin dicellui livre. Pareillement je prengz desdis acteurs le commencement du Cinquiesme livre ensievant Jusques la ou il dist. Au mi destroit de la bataille illec mist en pieces etc. Si ne lai pas seulement translate mais

41 Miniature from Valerius Maximus, *Faits et dits mémorables des romains*
Bruges, circa 1475–1480

17.5 x 19.4 cm (6 15/16 x 7 5/8 in.)
Ms. 43; 91.MS.81

Plate: Master of the Dresden Prayer Book, *The Temperate and the Intemperate*

The Memorable Deeds and Sayings of the Romans is a compilation of stories about ancient customs and heroes. Written in the first century A.D. by Valerius Maximus (fl. circa A.D. 20), it continued to be read during the Middle Ages. Loosely organized by moral and philosophical categories (temperance, charity, cruelty, etc.), Valerius Maximus, as the book is often called, served as a textbook of rhetorical exercises. Its popularity grew in the later Middle Ages due to vernacular translations, such as the French one commissioned by Charles V of France (r. 1364–1380). This cutting derives from a folio-size copy of the French translation made for Jan Crabbe, the Abbot of the Cistercian Abbey at Duinen, south of Bruges.

This large miniature appeared at the beginning of book 2. It shows Valerius instructing the emperor Tiberius (to whom he dedicated the text) on the value of temperance. In a spacious dining room, the upper classes shown at the back behave decorously—displaying temperance—while in the foreground the antics of lower-class characters illustrate the antithesis. In the hands of the prolific Master of the Dresden Prayer Book, a witty anonymous illuminator from Bruges, the appropriate behavior seems staid, while the bad example amuses us. Over the next two centuries drunkenness and other foibles of the middle and lower classes would become beloved and even trademark subjects of Flemish painters. They were preceded by Flemish illuminators, who left us a trove of miniatures of social customs and behavior that inform us about the values of the time. TK

42 Miniature from a Book of Hours
Probably Ghent, before 1483

12.5 x 9 cm ($4\frac{7}{8}$ x $3\frac{1}{2}$ in.)
Ms. 60; 95.ML.53

Plate: Attributed to the Master of Mary of Burgundy, *The Annunciation to the Shepherds*

One of the geniuses of the Golden Age of Flemish painting in the fifteenth century is the enigmatic manuscript illuminator named the Master of Mary of Burgundy. He takes his name from Mary, Duchess of Burgundy (1457–1482), among the most powerful of his presumed patrons. He practiced his art only from around 1470 to 1490, and he worked in the region of Ghent in Flanders, where he was an associate of Hugo van der Goes (circa 1436–1482), whose paintings strongly influenced him. The Master of Mary of Burgundy was the only Flemish artist of the time whose work rivaled van der Goes's works in their emotional power and their sympathy for common people. In this miniature the shepherds have the coarse and rugged features of peasants in paintings by van der Goes. Their faces are drawn with a richness of modeling and precision of contour that find no equal in Flemish manuscript illumination. The artist's achievement is all the more remarkable when one considers that he customarily painted in this very small format.

The nocturnal scene with rolling hills is lit only by the glow of the graceful angel high in the sky, by a diminutive ballet of gilded angels gliding down toward the manger, and by the light within the stable itself. Nocturnal subjects strongly attracted Flemish, Dutch, and French painters in the last quarter of the fifteenth century.

The miniature probably comes from an elaborate illuminated book of hours that is now in the Houghton Library at Harvard University, Cambridge, Massachusetts. That book's decoration represented a collaboration with Simon Marmion (see no. 35) and the Master of the Dresden Prayer Book (see no. 41), two of the other leading artists of the day. It was possibly made for a Spanish patron. Unfortunately, many of the book's other full-page miniatures are lost. TK

Comment le duc de berry fist
les nopces de son filz et de sa
fille madame marie de france
et du filz au conte de blois. Et
dudit duc et de son filz qui fu
rent veufues. Et comment le
conte destampes fut envoye de
vers le duc de bourgongne po[ur]
traicter a lui de la paix et les
responces que le duc lui fist.
Chapitre. iiii.xx. v.

En lan de lin
carnation nre
seigneur mil
troiscens iiii.xx.
et six ou mois
daoust se de
party le conte guy de blois z la
contesse marie sa femme bien
acompaignies de chevalliers
et descuiers de dames et de
damoyselles z en bon arroy et
bien ordonne de la ville de blois
et se mirent au chemin pour
venir en berry et emmenerent
avecques eulx leur jeune filz
qui lannee devant avoit fiancee
marie fille au duc jehan de berry
et estoit lintencion au conte de
blois et a la contesse que eulx
venuz a bourges en berry leur
filz procederoit avant ou mari
age z aussi estoit telle linten
cion du duc de berry et a la du
chesse sa femme. Si que quant
toutes ces parties furent venues
les unes devant les autres le
mariage de ces deux jeunes

43 Jean Froissart, *Chroniques*, Book 3
Bruges, circa 1480

366 leaves, 48.2 x 35 cm (19 x 13¾ in.)
Ms. Ludwig XIII 7; 83.MP.150

Plate: Master of the White Inscriptions, *The Marriage of Louis de Blois and Marie de France*, fol. 288v

The monumental *Chronicle* written by Jean Froissart (1337–circa 1410), covering the period from around 1322 to 1410, is the most famous historical record of the fourteenth century. It recounts the major political and military events of the time, focusing on the rivalry between England and France. The *Chronicle* is a basic resource for the study of the Hundred Years' War (circa 1337–1453), the ongoing conflict between these kingdoms. Froissart also describes the affairs of other realms, though largely as they relate to the complex network of overlapping and shifting alliances around the protagonists. The Getty manuscript contains only book 3 (of four), which describes "the recent wars in France, England, Spain, Portugal, Naples, and Rome." Its 730 pages cover the period from 1385 to 1389, an indication of the level of detail Froissart sought to impart. He conducted plenty of research. For book 3 he traveled to the territories ruled by the Count of Foix in southwestern France to gather information on events in the region and on the Iberian peninsula.

The Getty volume shows the lasting esteem that the *Chronicle* enjoyed. It was produced about seventy years after the author's death, when a number of other copies of his *Chronicle* were transcribed and illuminated. This one was painted in Flanders, perhaps in Bruges. The choice of subjects for the sixty-four miniatures strongly emphasizes events involving the English, evidence perhaps that the book was produced for the insular market. The English and the Burgundians, who ruled from various towns in Flanders and northern France, were allies during this period, and the English exhibited a strong taste for all things Burgundian, including Flemish paintings, tapestries, and illuminated manuscripts. Margaret of York, Duchess of Burgundy (1446–1477), helped her brother, the English king Edward IV (r. 1461–1483) to obtain various books, tapestries, and other treasures from Flemish artists. Circumstantial evidence suggests that Edward himself may have purchased this book for his library.

The miniature reproduced here illustrates the marriage of Louis de Blois and Marie de France, the daughter of the Duke of Berry, at the portal of the Cathedral of Saint Étienne in Beauvais in 1386. Louis was the son of Froissart's patron Guy, Count of Blois, who commissioned book 3 of the *Chronicle*. Consistent with the artistic tradition of the time, the wedding couple and their party wear the extravagant fashions of the Burgundian court in the illuminator's day—not the costumes of the fourteenth century.

TK

I·N·RI

44 Miniature, Perhaps from a Manuscript
Probably Franconia (Germany), last quarter of the fifteenth century

38.8 x 24.3 cm ($15\frac{5}{16}$ x $9\frac{9}{16}$ in.)
Ms. 52; 93.MS.37

Plate: *The Crucifixion*

This monumental *Crucifixion* shows the deceased Christ on the cross on the mount of Calvary. Three angels—two at his wrists, another at his feet—capture his blood in chalices. Below, the sorrowful mother of Jesus lowers her head, her eyes closed and her hands crossed over her bosom. Opposite her, John the Evangelist stands quietly, his right arm reaching over his heart. The ritual symbolism in this representation of the Crucifixion was popular in Germany around 1500. The capturing of the blood of Christ in the chalice of Holy Communion refers to the transubstantiation of bread and wine in the Eucharist, the sacrament celebrated in the mass. In receiving Communion worshipers partake of bread and wine that has been consecrated by the priest at the altar. The body and blood of Christ are understood to be present in the Eucharistic elements.

The miniature's background shows Jerusalem in the guise of a prosperous German town of the late fifteenth century. Although not identified securely, its location on a sloping bluff and with a river passing through it may be inspired by the topography of the bustling metropolis of Nuremberg in Franconia. The skull and bones at the foot of the cross allude to "Golgotha," the Hebrew name for Calvary, meaning "place of the skull." The skull may also refer to Adam, who was thought to be buried there.

A full-page Crucifixion miniature is the most important illustration in a missal (or mass book) and often the only one. It is located at the canon of the mass, the Communion prayer. A number of missals printed in Germany at this time have woodcut illustrations for the canon with similar allusions to the celebration of the Eucharist. In these representations, as here, the cross in the shape of a T derives from a long-standing tradition in which it also served as the first letter of the canon, which begins *Te igitur clementissime pater* (You, therefore, most merciful father). This miniature may thus have been painted for inclusion in such a missal. If so, the book would have been exceptionally large and impressive. TK

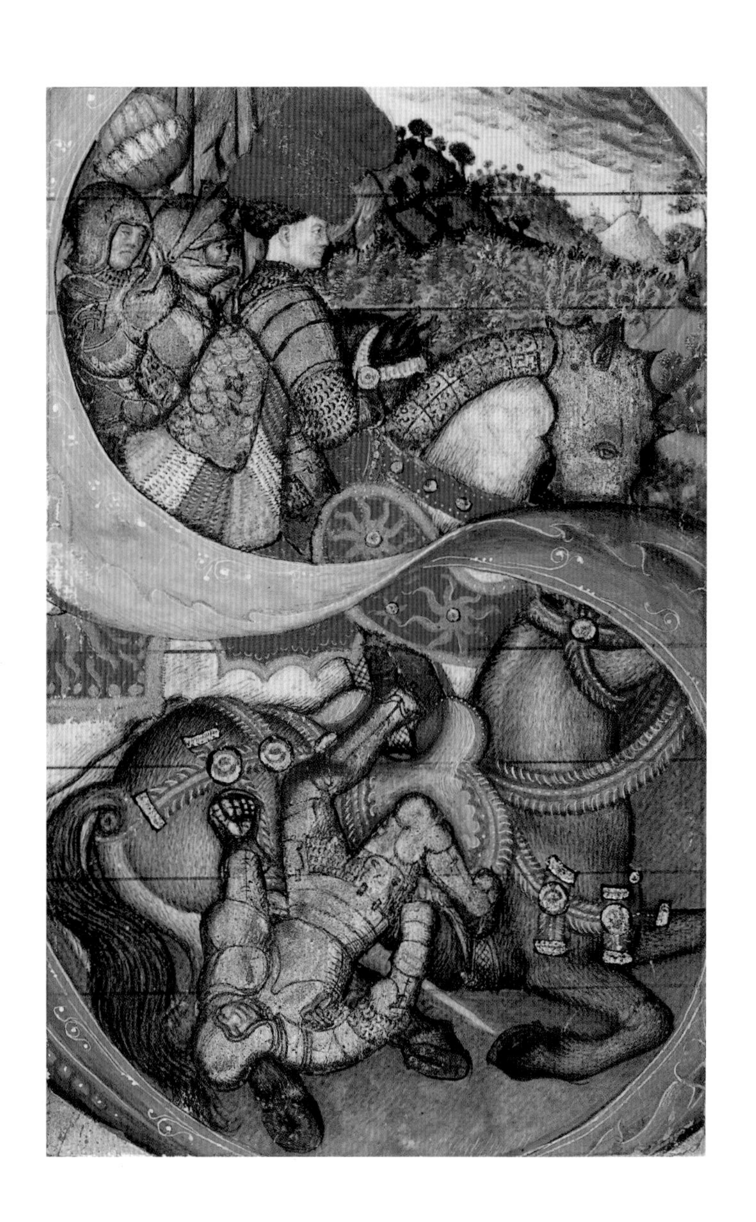

45 Historiated Initial from a Gradual
Probably the Veneto, possibly Verona, circa 1440–1450

14.2 x 9 cm (5⁹⁄₁₆ x 3½ in.)
Ms. 41; 91.MS.5

Plate: Attributed to Antonio Pisano, called Pisanello, and the Master of Antiphonal Q of San Giorgio Maggiore, Initial *S* with *The Conversion of Saint Paul*

Saint Paul, one of the most significant figures in the formation of the Catholic Church, endeavored to spread the Gospel beyond the Jews to the world at large. This historiated initial *S* (now closely trimmed) illustrated the mass for the feast of the Conversion of Saint Paul (January 25) in a gradual, a book containing the chants sung during mass. While journeying to Damascus, Saul, a Jew, experiences a light from heaven that engulfs him and his companions (Acts 9:1–9 and 26:12–18). He falls to the ground and hears the voice of Jesus calling him to become a Christian evangelist. Saul converts, changes his name to Paul, and preaches the new faith.

In this initial, Saul, dressed in a soldier's helmet and armor, tumbles to earth, his horse collapsing beneath him. Unexpected in this otherwise characteristic representation of Saul's conversion is his unassuming presence, his face barely visible beneath the helmet. The illuminator focuses instead on another soldier, seated erect on a more elaborately liveried steed in the top half of the initial. His tall, fancy *cappuccio* (hat) and the *giornea* (tunic) trimmed in green, white, and red—the colors of both the Gonzaga and Este families—suggest that he is the leader of this band of Italian soldiers. The noble and contemporary costume and the vivid profile suggest that he is not a biblical figure at all, but a youthful scion of one of these ducal families. He may have been the book's patron, commissioning the gradual for his family's use or for an ecclesiastical foundation under his protection.

Both the Gonzaga of Mantua and the Este of Ferrara were patrons of Pisanello (circa 1399–1455). Medalist, fresco painter, painter on panel, portraitist, history painter, and possibly a manuscript illuminator too, this versatile artist moved among the courts of northern Italy, Rome, and Naples. Although not all scholars consider this initial to be painted by him, Pisanello's originality and descriptive powers are evident here in the expressive use of silver to convey the luster of armor, the splendor of the central figure's silhouette including his delicate facial features, and the powerful rendering of Paul's horse. Pisanello's depiction of horses, with their muscular haunches, are among the more memorable ones in European art.

The initial's landscape was painted by an anonymous artist who worked in Verona.

TK

46 Miniature from a Devotional or Liturgical Manuscript
Possibly Mantua,
circa 1460–1470

20.1 x 12.9 cm (7 15/16 x 5 1/16 in.)
Ms. 55; 94.MS.13

Plate: Girolamo da Cremona, *Pentecost*

Italian artists of the fifteenth century applied mathematical principles in designing a painting. These Renaissance rules of composition—frequently adapted and rethought—would have an impact on European painting continuously down to our own day. In this miniature of Pentecost, the descent of the Holy Spirit upon the apostles, nearly all the elements are arranged symmetrically around an invisible central axis. The solemn, columnar figure of the Virgin Mary along with the Holy Spirit in the form of a radiant descending dove indicate this axis. Equidistant from the axis appear the windows, the portals in the flanking walls, the pair of candlesticks on the mantel, and the apostles themselves. The two groups of apostles are organized in a mirror image of one another, a back row of three, a middle one of two, and in the foreground one each. The kneeling figures open around the Virgin like a pair of wings welcoming us. We look over the shoulders of the foremost apostles to participate.

The artist avoids the monotony of strict symmetry by varying details, such as the colors of the apostles' robes, the men's gestures, and the arrangement of books around the candles, and by showing an open window with a view opposite a closed one. The geometric clarity of this design, the Virgin's imposing height, and the tall proportions of the room give this scene a monumental quality, even though the miniature itself measures only eight inches from top to bottom.

The illuminator is Girolamo da Cremona (fl. 1458–1483), a protégé of the great painter Andrea Mantegna (circa 1431–1506). Girolamo moved among the powerful courts of Northern Italy. He illuminated books in Ferrara, Mantua, Siena, and Venice. Besides the thoughtful composition, another pleasure of Girolamo's art is his ability to describe the different textures of materials, from the stone window frames and the window's bull's-eye glass to the dull red bricks of the walls and the dyed leather bookbindings.

The *Pentecost* was made for a liturgical book or a book of private devotion. No other part of the manuscript has come to light. TK

47 Gualenghi-d'Este Hours
Ferrara, circa 1469

211 leaves, 10.8 x 7.9 cm
(4¼ x 3⅛ in.)
Ms. Ludwig IX 13; 83.ML.109

Plates: Taddeo Crivelli,
Saint Gregory the Great, fol. 172v
Saint Catherine, fol. 187v
Saint Bellinus Receiving the Gualenghi Family at the Altar, fol. 199v
Saint Anthony Abbot, fol. 204v

Devotion to the saints was one of the most popular aspects of Christian piety throughout the Middle Ages and Renaissance. Saints served as intermediaries between heaven and earth and performed miracles of faith and healing for the devoted. They were petitioned by the faithful, who viewed them as special advocates before God. The virtuous lives and deeds of these holy men and women were also looked to by ordinary people as examples to be followed in their own lives. In the visual arts the cult of saints was expressed in the reliquaries and churches built to house their earthly remains, in the illustrated books devoted to their legends (see no. 23), and in their numerous representations in sculpture and painting. Devotion to saints was also an integral part of many books of hours, which contained short prayers to the saints, often illustrated, in the section of the book known as suffrages.

In this book of hours, created for Andrea Gualengo (d. 1480) and his wife, Orsina d'Este, the majority of the figural decoration is devoted to the suffrages. Andrea came from a family of high-ranking courtiers at the Este court in Ferrara and himself held important advisory and ambassadorial posts during the reigns of Borso d'Este

(r. 1450–1471) and Ercole d'Este (r. 1471–1505). The Gualenghi family is depicted in the miniature accompanying the prayer to Saint Bellinus (fol. 199v), a twelfth-century bishop of Padua who must have held a special importance for the patron. The painting illustrates explicitly the relationship among mortal, saint, and God and underlines the intermediary role of the saint. The family kneels in prayer before the altar at which Saint Bellinus is celebrating mass. With one hand the saint clasps the outstretched arms of Andrea Gualengo while he gestures toward heaven with the other.

Saint Gregory the Great (circa 540–604) is also shown in an act of devotion directed toward heaven (fol. 172v). Seated before an altar, he looks up toward the divine light entering the niche overhead and opens his mouth as if in song. As in many other paintings in this book, Taddeo Crivelli infused the subject matter of divine presence entering into the world with a sense of spiritual rapture; the putto tangled in a scroll, the twisting blue banderole, and the sharp shimmering rays of gold in the border as well as the energetic lines of the marble behind the saint's head endow the painting with a heightened emotional pitch.

KB

48 Gradual
Rome, late fifteenth or early sixteenth century

188 leaves, 64.1 x 43.5 cm
(25 5/16 x 17 1/8 in.)
Ms. Ludwig VI 3; 83.MH.86

Plate: Antonio da Monza,
Initial *R* with *The Resurrection,* fol. 16

Toward the end of the fifteenth century, artists started exploring the newly discovered Golden House of Nero (an ancient imperial villa in the city of Rome) in order to study the walls of its rooms, which were covered with painted and stucco ornament. Visitors to the site were captivated by the fantastic creatures, candelabra, garlands, and delicate architectural elements represented on the interior walls. The Renaissance mania for all things of the classical world meant that the motifs, known as grotesques because of their association with the underground "grottoes" of the unexcavated house, were rapidly incorporated into the ornamental vocabulary of High Renaissance painting.

Fra Antonio da Monza, the illuminator of this large gradual made for the Franciscan Church of Santa Maria in Aracoeli in Rome, was one of the many Italian Renaissance artists who were profoundly influenced by the remains of classical art. The painted embellishment of the Getty's gradual not only draws on the sort of motifs found on the walls of the Golden House of Nero but also includes representations of antique cameos and other gems.

The opening page of the mass for Easter Sunday (fol. 16) is the most elaborate in the book, and its illumination is a stunning accomplishment of decoration *all'antica* (in the antique manner). The Christian subjects are the Resurrection (in the initial *R*), the Martyrdom of Saint Sebastian (seen through a glass cylinder that forms part of the letter *R*), the Annunciation (in a pair of roundels in the border on the sides of the page), and a bust of Christ (in the lower border). This Christian imagery shares the page with a wealth of classically inspired hybrid creatures and putti, all presented within a composition reminiscent of the schemes on the walls of Roman imperial houses. ECT

Esurre
xi ⁊ ad
huc tecu sum alle lu ya
posui sti super me manu
tu as alle lu ya mira
S·P·Q·R

INCIPIT ARGVMENTVM IN EPISTOLAM BEATI PAVLI APOSTOLI AD ROMANOS

ROMANI ſunt partis Italiæ. Hi præuenti ſunt a falſis Apoſtolis et ſub nomine domini noſtri Ieſu Chriſti in legem et prophetas erant inducti. Hos reuocat Apoſtolus ad verā et euangelicam fidem: ſcribens eis a Corintho

INCIPIT EPISTOLA BEATI PAVLI APOSTOLI AD ROMANOS

CAPITVLVM PRIMVM

PAVLVS ſeruus Ieſu Chriſti vocatus Apoſtolus: ſegregatus in euangelium dei quod ante promiſerat per prophetas ſuos in ſcripturis ſanctis de filio ſuo, qui factus eſt ei ex ſemine Dauid ſecundum carnem. Qui prædeſtinatus eſt filius dei in virtute ſecundum ſpiritum ſanctificationis ex reſurrectione mortuorum Ieſu Chriſti domini noſtri: per quem accepimus gratiam et Apo

49 Getty Epistles
France, circa 1520–1530

112 leaves, 16.4 x 10.3 cm
(6½ x 4¹⁄₁₆ in.)
Ms. Ludwig I 15; 83.MA.64

Plates: Master of the Getty Epistles, *Saint Paul* and Text Page, fols. 5v–6

See pages 114–115

In both content and appearance this French book is a product of the Renaissance. It offers distinctive evidence of the diverse paths by which the rebirth of learning and the visual arts that began in Italy spread throughout Europe in the sixteenth century. Toward the beginning of the century scholars took up the study of Saint Paul's letters with renewed fervor. The humanist Erasmus of Rotterdam (circa 1466–1536) and other Church reformers were attracted by his teachings to the Romans. Their interpretation of the epistles as sanctioning justification through faith rather than deeds became a topic of theological debate.

The Master of the Getty Epistles was the leading artist in a popular workshop of the Loire valley that specialized in decorating devotional books during the 1520s. The sources of his art are complex. The figure of Saint Paul, muscular and swathed in heavy robes, ultimately takes its inspiration from the art of Michelangelo, but the artist, who was trained in Flanders, probably knew the art of the Italian master only through his Northern European followers. The meandering, hilly, and spacious setting reflects the nascent art of landscape painting that made the Flemish school in Antwerp celebrated at this time.

The border of fruit and flowers is also Flemish in inspiration, while the elaborate architectural border framing the miniature shows many elements from ancient Roman architecture recently revived in Italy. Also pointedly Italianate is the crisp and easily readable humanist script, itself a revival of medieval Carolingian letter forms that humanists believed to be ancient. Moreover, the separation of the rubrics (or headings) from the text, their symmetrical design, and the spaciousness in arranging the components reflect the new attitude toward page design found in Italian printing. In this two-page opening, the arrangement of the text has received as careful attention as the composition of the miniature. Thus diverse threads of the artistic, intellectual, and technological ferment of the Renaissance are interwoven on the pages of the Getty Epistles.

TK

50 Spinola Hours
Ghent or Mechelen,
circa 1510–1520

312 leaves, 23.2 x 16.6 cm
(9 1/8 x 6 9/16 in.)
Ms. Ludwig IX 18; 83.ML.114

Plates: Gerard Horenbout
(The Master of James IV of Scotland),
The Holy Trinity Enthroned and
Abraham and the Three Angels,
fols. 10v–11

See pages 118–119

The Spinola Hours (named for the Genoese noble family that once owned the book) is one of the most sophisticated Flemish manuscripts of the sixteenth century. It contains eighty-eight miniatures within six hundred pages. Every page that lacks a miniature has fully decorated borders, most painted illusionistically with flowers and insects. For all the spiritual gravity of their subject matter, the miniatures in this book are often playful, teasing the viewer to believe in their painted illusions.

Miniatures in books of hours generally appear above the opening words (or incipit) of the book's main devotions. In this manuscript the miniature, illustrating a set of devotions called the Hours of the Holy Trinity, appears not only above the text but beside and below it, filling those regions of the page where a painted border traditionally appeared. The Trinity is shown as three persons in one: God the Father, Jesus Christ, and the Holy Spirit. They hold an orb, the symbol for universal dominion, while the central figure raises his hand in blessing. To further challenge our perceptions, the incipit appears not only on the genuine parchment but also on a slip of parchment painted on the miniature. This piece of parchment is "pinned" to the flat surface of the otherwise spacious miniature so that one painted illusion reveals the other for what it is.

Each major opening in the book has two miniatures. On the page facing *The Holy Trinity Enthroned* appears the Old Testament story of the elderly Abraham offering hospitality to three angels. They have come to announce that Sarah, his old and barren wife, will bear a child (Genesis 18:1–19). In the foreground Abraham bows down to the angels when they first appear. Above, as Abraham offers them food, Sarah peeks out of the opening in the tent behind them, smiling at the surprising tidings. The three angels were viewed as an Old Testament prefiguration of the Holy Trinity.

Gerard Horenbout was the finest Flemish illuminator of the first two decades of the sixteenth century and court painter to Margaret of Austria, Regent of the Netherlands. Circumstantial evidence suggests that this ambitious and expensive book, which engaged the talents of a host of prominent illuminators, including Simon Bening (see nos. 51–52), may have been made for her. TK

Die dn̄ica. hore de scā tri
Domine nitate. Ad mat.
labia mea aperies.

Et os meum annuntia
bit laudem tuam.
Deus ĩ adiutoriuz

51 Prayer Book of Cardinal Albrecht of Brandenburg
Bruges, circa 1525–1530

337 leaves, 16.8 x 11.5 cm
(6⅝ x 4½ in.)
Ms. Ludwig IX 19; 83.ML.115

Plate: Simon Bening,
Christ Before Caiaphas, fol. 128v

The advent of the printed book, which was introduced in Europe in the mid-fifteenth century, did not make hand-written books obsolete for many generations. Indeed, the text of this manuscript, a series of prayers relating to the Passion of Christ, is copied from a book printed in Augsburg in 1521. Cardinal Albrecht of Brandenburg, Elector and Archbishop of Mainz, wanted a hand-written copy of the printed book on vellum, illustrated with woodcuts. He then hired the illuminator Simon Bening to supply a series of forty-two full-page miniatures (along with historiated borders and other decorations). Albrecht probably preferred not only Bening's artistry to that of the woodcut illustrator, but the luxuriousness and durability of parchment to paper and the saturated colors of illumination to the black-and-white of the woodcut. Competition with the printed illustration probably spurred the unrivaled period of creativity and originality that characterizes Flemish illumination after 1450.

Here the combination of the verisimilitude of Bening's art with a great story told in many scenes results in an uncommonly vivid and moving narration. The artist exploits the drama inherent in turning the page, so that each turn reveals a new confrontation between Christ and his persecutors. Through the accumulation of narrative incident and subtleties of characterization, Bening's Christ comes alive. The artist underscores his human side and vulnerability, encouraging the reader to identify with his suffering.

Bening further heightens the drama with the nocturnal setting; many scenes are illuminated, as here, only by torchlight. This scene shows Christ following his betrayal in the Garden of Gethsemane being led before the High Priest Caiaphas. Caiaphas tears his own robes and calls Jesus a blasphemer when he identifies himself as the Messiah. Bening suggests Christ's divinity in his impassive acceptance of his destiny and his physical beauty.

Archbishop Albrecht was a true Renaissance prince in his love of art, learning, and luxury. He commissioned another book from Bening and paintings or graphic arts from the leading German masters Dürer, Grünewald, and Cranach. TK

52 Miniature from a Book of Hours
Probably Bruges,
circa 1540–1550

5.6 x 9.6 cm (2³⁄₁₆ x 3³⁄₄ in.)
Ms. 50; 93.MS.19

Plate: Attributed to Simon Bening, *Gathering Twigs*

Since the era of the Renaissance, landscape painting has attracted artists and collectors alike. Its appeal, from such masters as Pieter Brueghel the Elder (1525/30–1569) to Claude Monet (1840–1926), is broad. Landscape painting remains one of the most popular attractions of modern museums. The European tradition of landscape painting sprang from a variety of sources, one of the most original and important being the calendar illustrations of late medieval devotional books. Since antiquity the months and seasons of the year were important subjects in art. Artists represented the months symbolically by the zodiacal signs and with figures performing the agricultural labor associated with a particular month, such as sowing or harvesting. In the fifteenth century, book painters showed that illuminations of the settings where the workers toiled, with their distinctive weather conditions, could be even more evocative of a particular month than the labors themselves.

This cutting painted by Simon Bening of Bruges (1483/84–1561), illustrates the gathering of twigs, the "labor" for one of the winter months. It appeared originally in a book of hours in the lower border (called *bas-de-page*) of the page for February in the book's calendar of Church feasts. It shows a damp but sunny winter day. The artist engages our eye not only in the tactile details of the foreground but in the palpable atmosphere that draws us to the middle distance and the gently rolling hills beyond. This diminutive scene is as ambitious in scope and composition as independent paintings of considerably larger dimensions. It is therefore not surprising that the cutting's previous owner admired it as such. Despite its size, he had it framed and hung it on the wall like any other landscape painted on canvas or wood. TK

53 *Mira calligraphiae monumenta*
Vienna, 1561–1562 and
circa 1591–1596

150 leaves, 16.6 x 12.4 cm
(6 9/16 x 4 7/8 in.)
Ms. 20; 86.MV.527

Plates: Joris Hoefnagel,
A Sloth (?), fol. 106
How to Construct Lowercase f and g,
fol. 143v

See pages 124–125

During the sixteenth century elaborate and inventive calligraphy, or display script, was admired in humanist circles. Intellectuals valued the inventiveness of scribes and the aesthetic qualities of writing. In 1561 and 1562 Georg Bocskay, the Croatian-born court secretary of the Holy Roman Emperor Ferdinand I in Vienna, created this *Model Book of Calligraphy* to demonstrate his unrivaled technical mastery of the immense range of writing styles known to him. He arranged the calligraphy cleverly, giving each page of the book an independent beauty. Indeed, this model book appears not to have been intended originally for painted decoration (even though some pages are written in gold and silver). About thirty years later Joris Hoefnagel, who became a court artist of Ferdinand's grandson, Rudolf II, was asked to illuminate the book. He added fruit and flowers to nearly every page, composing them so as to enhance the unity and balance of the already written pages. The result is one of the most unusual collaborations between scribe and painter in the history of manuscript illumination.

The Antwerp-born Hoefnagel illuminated only six manuscripts, although each was as elaborate as the Getty book and one is said to have required eight years to complete. He also produced countless watercolors of *naturalia,* along with landscapes and city views. He is thus recognized as an influential figure in the emergence of Netherlandish still-life painting in the seventeenth century.

Hoefnagel added to the back of the *Model Book of Calligraphy* some intricately designed pages that instruct the student in the art of constructing the letters of the alphabet in upper- and lowercase. This section has broader, more complex imagery that addresses intellectual and political interests of the court of Rudolf II in Prague. Laden with symbolism, it contains many references to the emperor himself. TK

BEATI ILLI, QVI SVBVENI-
VNT MISERIS: QVONIAM E-
IS REPENDITVR, VT PER
MISERICORDIAM DEI DE MI-
SERIIS LIBERENTVR: NAM
IDIPSVM VIDETVR IVSTVM
VT QVI À POTENTIORE A-
DIVVARI VVLT, ADIV-
VET INFERIOREM IN
QVO IPSE EST POTENTIOR
MISERICORDIA LARGIOR, VBI FIDES PROMPTIOR. NIHIL
TAM COMMENDAT CHRISTIANVM, QVAM MISERATIO CHAR-
ITATIS. ANNO DOMINI. M.D.LXI.

f
g

GLOSSARY

Apocrypha The Old Testament apocrypha are sacred writings included in the Greek and Latin Bible but not in Hebrew scripture. New Testament apocrypha are early Christian writings proposed but not accepted as part of the Bible.

Cadelle A capital letter flourished with wide, parallel pen strokes with occasional cross strokes.

Codex A bound manuscript volume.

Decorated initial An enlarged, painted letter embellished with non-figural decoration.

Divine office The prayer liturgy of the Catholic church, consisting primarily of the recitation of psalms and the reading of lessons; divided into eight daily services: Matins, Lauds, Prime, Terce, Sext, None, Vespers, and Compline. The office is recited daily by monks, nuns, and clerics.

Drollery An amusing or whimsical figure. Drolleries include hybrid figures and usually appear in the margins of manuscripts.

Evangelist One of the authors of the four Gospel accounts of Christ's life: Saints Matthew, Mark, Luke, and John.

Extender A decorative enhancement of an initial that continues the letter form into the margin.

Folio A manuscript leaf. The front side is called the *recto* and the back the *verso*.

Historiated initial An enlarged, painted letter that contains a narrative scene or identifiable figures.

Humanism A cultural and intellectual movement inspired in part by the revival of classical learning in the Renaissance.

Icon The Greek work for "image." In Byzantine culture, an icon (most often in the form of a small painting on panel) carries the likeness of a sacred person or subject to be venerated.

Iconography The subject matter of an image; also, the study of the meaning of images.

Incipit The opening words of a text. An incipit page is an elaborately decorated page that introduces a section of text.

Inhabited initial An enlarged, painted initial containing human figures or animals that cannot be identified specifically.

Laity The Christian faithful who are not monks, nuns, friars, or members of the clergy.

Liturgy Public religious ritual.

Mandorla An almond-shaped aureole surrounding the body of a deity or holy figure.

Mass The Christian service focused on the sacrament of the Eucharist, in which bread and wine are consecrated and shared.

Miniature An independent, framed illustration in a manuscript.

Order A group of people living under a religious rule.

Paleography The study of historical scripts.

Palette The range, quality, or use of color.

Parchment or vellum Prepared animal skin commonly used as the writing surface in manuscripts of the Middle Ages and Renaissance.

Passion of Christ The sufferings of Jesus leading up to and including the Crucifixion.

Putto (pl. putti) A nude infant, often with wings.

Scriptorium A room for the writing of texts; also, a group of people working together to produce manuscripts.

Vernacular The spoken language of a region, such as French or German, as opposed to an international language, such as Latin or Greek. During the course of the Middle Ages, literature came to be written in the European vernacular languages.

INDEX

Numerals refer to page numbers